Calculus of Life

Sandeep Chavan

Published by SANDEEP CHAVAN, 2025.

CALCULUS OF LIFE

First edition. July 15, 2025.

ISBN: 979-8231625727

Written by Sandeep Chavan.

Table of Contents

To all the curious souls who fell in love with mathematics—not for marks, but for meaning.And to those who *cleverly* used its elegance to veil the deeper truths of nature...May your equations one day ripple into alignment.

Ha ha. With affection,— *Sandeep*

"We are not constants.We are functions of time,differentiated by experience,integrated by memory,and aligned—only when the ripple meets its field."

— *Sandeep Chavan*

Disclaimer

This book is not a mathematics textbook—nor is it a spiritual manual, scientific doctrine, or psychological guide. It is a reflective exploration using the language of calculus to interpret life, thought, emotion, and existence. The analogies, metaphors, and concepts presented herein are intended to inspire insight, not to provide academic instruction or therapeutic advice.

While key ideas from **calculus** and **Universal Energy Dynamics (UED)** are referenced, they are expressed in a simplified, metaphorical style to resonate with a wide audience. Readers seeking rigorous mathematical or scientific formulations are encouraged to consult formal textbooks and research papers.

Any resemblance to real-life experiences, institutions, or individuals—outside the author's own story—is purely coincidental unless explicitly mentioned. This book invites open-minded inquiry and personal reflection; it is not meant to replace formal education or scientific consensus.

The author and publisher disclaim any liability arising from the use or interpretation of this work outside its intended context—as a philosophical and metaphorical journey.

Preface

They say mathematics is the language of the universe. I say—it's the echo of life itself.

This book was not planned in a lab or outlined in a publishing room. It was born in a classroom, somewhere between a chalk-smeared blackboard and a student's confused gaze. It grew through decades of teaching Physics, Chemistry, and Mathematics, and through my personal journey of failures, realizations, misalignments, and rediscovered intent. What you now hold is not a textbook. It is a lived experience—seen through the lens of calculus, but felt through the pulse of life.

Why calculus?

Because no other branch of mathematics captures change, growth, limit, and consequence as intimately as calculus. It isn't just about solving problems—it's about observing them unfold, moment by moment. Life, like a function, doesn't remain constant. It accelerates, it decelerates, it breaks, it heals. It has turning points, inflection points, asymptotes, and even undefined spots. Just like us.

In this book, I've tried to strip away the abstraction and reintroduce calculus as a philosophy—one that belongs not just in math classrooms, but in boardrooms, meditation halls, operating tables, artist studios, and kitchen sinks. This is calculus for the thinker, the teacher, the student, the scientist, and above all, for the human.

Underneath it all lies the silent field I call **Universal Energy Dynamics (UED)**—the philosophical and scientific backbone of every ripple, derivative, and integral discussed here. You don't need to understand UED to grasp this book. But once you read it, you may begin to see your life—and the universe—differently.

This book is not for solving equations. It is for dissolving illusions.

Read slowly. Reflect deeply. Reread when life changes again.

And remember: the integral of your journey is still being written.

— *Sandeep Chavan*

Prologue

"*What if life was never random—but simply unresolved?*"

We grow up learning calculus in school as a branch of mathematics—an abstract toolkit to solve problems about motion, force, growth, decay, area, and change. We solve for velocity. We find slopes. We calculate maxima. But few of us ever stop to ask: *Is this only math? Or is it a reflection of how life itself unfolds?*

This book does not teach calculus. It reveals that *calculus has always been teaching us.*

We live in a world obsessed with results. Exams, outputs, deliverables, rankings. We celebrate those who integrate more—achieve more—and those who differentiate better—react faster. But beneath all these performances, life is happening not as a set of numbers, but as a ripple—one consequence after another, a dynamic interaction between our intent and our surroundings.

Life is not linear. It's a function. A function of memory, emotions, environment, actions, and choices. And just like any function—it has limits, discontinuities, turning points, and sometimes, it diverges.

This book, *Calculus of Life*, is born from years of teaching science, engineering, and mathematics—not just as subjects, but as languages of reality. It brings together reflections from classrooms, coffee cups, collapsed dreams, philosophical storms, and scientific revolutions—tied together by one unifying thought:

"Every ripple we create—whether through thought, action, or emotion—has a slope, a curvature, and an area. And until we understand the shape of our ripple, we cannot resolve the meaning of our journey."

You will not find dense equations here. But you will find clarity. Each chapter is a simple reflection—rooted in the core ideas of calculus, retold in human terms. It is for the student who wonders what this math means. It is for the teacher who wants to teach beyond formulas. It is for the engineer who questions purpose. It is for the scientist who senses the cosmos is whispering something deeper.

And it is for you—who have lived all your life differentiating your pain, integrating your joys, and trying to find the constants that still remain when everything else changes.

So turn the page.

Let us not solve life. Let us integrate it—together.

Part I – Foundations of Change

1: The Living Function

You Are Not a Constant

At the heart of calculus lies a truth that quietly mirrors life: nothing alive remains constant.

We are taught in school that a constant is something that never changes—like the number 5, or the gravitational constant, or the speed of light in a vacuum. Constants are safe, predictable, unchanging. But life? Life is not a constant. Life is a function.

You are a function.

The Illusion of Stability

We often define ourselves with static labels:

- "I'm an introvert."
- "I'm not good at math."
- "I've always been like this."

These statements create the illusion that we are constants—fixed identities with no evolution. But in reality, every moment of living changes us. Our thoughts shift. Our emotions rise and fall. Our priorities evolve. Even our cells are replaced continuously.

To say "I am who I am" is mathematically equivalent to saying:

$$f(x) = c$$

A flat, lifeless line.

But if you've ever loved, lost, doubted, failed, tried again, or simply paused and changed your mind—you've proven:

$$f(x) \neq \text{constant}$$

You are living. Therefore, you are dynamic.

The Signature of a Function

In mathematics, a function is a relationship—a rule that assigns a unique output to every input. In life, your mind is that rule. Every situation you encounter, every environment you enter, every relationship you form—produces a different version of you.

You laugh with some people. You withdraw around others. You become sharp under pressure. You soften when safe.

You aren't being fake. You are functioning.

That's what a function does—it responds to input. And the beauty is: no matter the complexity of your inputs, you remain a valid function, even when misunderstood by others.

Being Misread ≠ Being Wrong

In calculus, we often don't see the entire function. We see data points, derivatives, small slices of behavior. The function may look wild or inconsistent over a certain interval—but that doesn't mean it lacks order. It may just need more domain to be understood.

Humans are like that too.

People judge based on partial inputs. A moment of anger, a year of failure, a bad decision. But your function is bigger than any local dip or spike. It's defined over a much broader domain than what anyone else sees.

You are not your lowest point. You are not your highest either. You are the ripple that connects both—and everything in between.

Constants Can't Feel Wonder

A constant doesn't adapt. A constant doesn't fall in love. A constant doesn't feel wonder at a starlit sky or fear when a loved one is ill. It doesn't question existence or discover new beliefs. But you do. Because:

You are a living function—capable of change, reflection, and recursion.

Even when you feel stuck, it's not because you are constant. It's because your function is waiting for the next input, the next variable shift, the next ripple to align and move you again.

The Value of Being Variable

We often crave stability. But stability doesn't mean stillness. In nature, even balance is a dynamic dance—ripples counterbalancing one another.

- A tree sways with the wind but doesn't break.
- A heart adjusts its rhythm to emotion.
- The earth spins and orbits, yet life persists.

So too, your value lies not in being fixed—but in being adaptable, aligned, and aware. A function that adjusts, evolves, and learns is far more powerful than any constant.

You Are Not a Constant. And That's Beautiful.

Mathematically, if you were constant, your derivative would be zero:

$$\frac{d}{dx}(c) = 0$$

That means:

- No slope.
- No growth.
- No ripple.
- No life.

But you're not zero. You are motion. You are consequence in motion. You are ripple intelligence resolving itself.

Reflective Questions

- What label have I accepted about myself that assumes I'm a constant?
- Where in life have I changed, even if I didn't realize it?
- Am I allowing myself the freedom to function—not just exist?

Closing Thought

"You are not a number to be memorized. You are a ripple equation, still revealing its truth."

As we move forward, remember this: life doesn't demand you be perfect. It asks that you keep functioning—with awareness, with honesty, and with grace.

Because only those who accept they are not constants...can ever discover the beauty of true growth.

2: Differentiation: The Art of Now

Measuring How We Change, Moment by Moment

In life, we often ask: "Am I improving?" "Why am I feeling this way?" "What's changing in me?"

These questions are not philosophical alone—they are differential. They reflect an instinct to understand rate of change, not just position.

In calculus, we define this idea formally. The derivative of a function tells us how that function is changing at a given instant. It's not the value of the function. It's the velocity of its becoming.

The Derivative Is the Present Moment

If a function is your journey, then the derivative is your *moment-to-moment* adaptation. It doesn't look back too far. It doesn't forecast far ahead. It zooms in—to a precise location on your life's curve—and asks:

"Right now, in this precise interval—what's happening to me?"

When you become aware of your emotions, your reactions, your state of mind in the now, you are practicing differentiation of the self.

And just like in math, this awareness gives you direction.

Slopes of Thought and Emotion

Think of it like this:

- When your mood is improving: your derivative is positive
- When you're declining mentally or emotionally: the slope is negative
- When you're in balance: slope = 0
- When things are erratic: the derivative is undefined or discontinuous

You may not always know why. But recognizing how fast and in what direction you're changing is the first act of self-awareness.

We've all had those moments:

"I don't know why, but I'm spiraling."

"Something shifted—I feel lighter now."

These aren't random feelings. They're differentials of your internal function.

Change Is Not the Enemy

Many people resist change because they fear instability. But differentiation is not instability—it's intelligence. Even in engineering and physics, systems are not feared for changing—they're feared when they change without control.

Life will change. You will change. The question is not *"Can I stop it?"* It is:

"Can I stay aware of the direction I'm changing in?"

That's what differentiation gives you.

Small Change, Big Insight

In math, we write:

$$\frac{dy}{dx}$$

Where:

- dy is the change in outcome
- dx is the change in input

In life:

- y could be happiness, stress, confidence, doubt
- x could be time, effort, relationships, decisions

So every choice you make—every thought you engage—produces a derivative response.

Sometimes, changing your input just a little—a walk, a breath, a kind word—can produce a significant change in your mental slope.

Small ripples. Large direction shifts.

Emotional Calculus

Let's try applying this directly.

Suppose your function is your emotional state over time.

Then:

- A sudden trauma = a sharp negative derivative
- A surprise compliment = instant slope upward
- Meditation = brings derivative closer to zero (emotional stillness)
- Reflection = helps find critical points (where growth

or decline pauses)

This is why mindfulness, therapy, writing, and even silence matter. They allow us to detect the derivative before the curve becomes uncontrollable.

The Art of Self-Differentiation

In relationships, "differentiation" also has a psychological meaning—it means the ability to stay connected to others without losing your own identity.

It's saying:

- "I love you, but I am still me."
- "I disagree, but I won't disconnect."

It's not emotional coldness. It's slope control. Keeping your own curve intact even as others try to influence it.

Where Differentiation Fails

In calculus, some functions have no derivative at certain points. The graph may be too sharp (a cusp) or too broken (a jump discontinuity).

So too in life:

- A moment of trauma may break your slope
- A sudden identity crisis may create a discontinuity
- Prolonged conflict may cause erratic behavior

And just as we do in calculus, we must treat such moments with care. Sometimes, it's not about calculating the slope. It's about acknowledging that the slope is undefined—and seeking a new approach, a new input, a new domain.

Practical Tool: "Life Derivative Check"

When feeling off, try asking:

1. What just changed? (dx)
2. How am I responding? (dy)
3. Is the direction helping or hurting? (Sign of derivative)
4. Can I slow this down or redirect it?

You don't need a full graph. You just need to know your slope.

Closing Reflection

"Differentiation is not about who you were yesterday. It is about *who you are becoming right now.*"

Every breath is a new point on your curve. Every choice, a new slope.

The art of living well lies in sensing the present shift—then realigning, recalibrating, and continuing with awareness.

You are not a fixed value. You are a curve in motion. And the present moment is your most honest derivative.

3: Limits: Where Life Begins to Turn

Personal thresholds and existential transitions

In mathematics, limits help us understand what happens when we approach a boundary—when we get infinitely close to a point, even if we never quite reach it. The limit is a threshold, a place where something begins to change, where one state ends and another is about to begin.

In life, we are constantly approaching such thresholds.

Some we see coming. Some arrive suddenly. And some, we never realize—until we've already crossed over.

This chapter is about those invisible edges where life begins to turn.

What Is a Limit?

Mathematically, the limit of a function as it approaches a point doesn't always tell you what the value *is*, but what the value *tends toward.*

For example:

$$\lim_{x \to a} f(x) = L$$

Means that as x gets closer to a, f(x)) gets closer to L.

Now replace that with:

- As you approach a decision, an age, a failure, a transition...
- Your response, your emotion, your belief—moves toward a new value.

That value is your limit.

The Limits of You

We all have personal limits:

- The most we can tolerate emotionally.
- The most pressure we can handle before breaking.
- The edge between certainty and doubt.
- The threshold where a belief collapses or a new one is born.

But unlike math, where we study limits from both sides, in life, we often live on only one side of the limit—never daring to approach it from another perspective.

And that's where misunderstanding grows.

Thresholds That Shape Us

There are certain moments in life where everything turns—quietly, irreversibly:

- The death of a loved one
- The birth of a child
- The day you quit your job
- The moment you decide to forgive
- The realization that you were wrong

- The encounter with a truth you can no longer deny

These are not conclusions. They are limits—existential turns where the function of your life shifts slope, redefines its domain, or becomes piecewise.

And sometimes, like in calculus, the function itself is not defined at the limit. But its approach tells the whole story.

One-Sided Limits and Emotional Bias

In mathematics, we check limits from the left and the right. But in human behavior, we often see only one side:

- We remember how someone hurt us (left-hand limit), but we don't see their reason or context (right-hand limit).
- We look at our failure from our ego's side (left), but not from the growth it generated (right).

True understanding of self and others requires both views. Otherwise, the function appears discontinuous—fragmented, flawed.

Jump Discontinuities: Sudden Life Changes

In calculus, some functions have jumps—places where the left limit and the right limit do not match.

In life, these are the moments when:

- You walk out of a relationship forever.
- You leave a city and never return.
- You stop believing in something overnight.

These are not errors in your function. They are jumps in your evolution. They remind us that not all transitions are smooth. Some are necessary leaps.

Approaching Infinity: The Existential Edge

What happens when we approach infinity?

In math, we say:

$$\lim_{x \to \infty} f(x)$$

We're trying to understand what happens at the edges of space, time, value.

In life, we also ask:

- What lies beyond this pain?
- What is the point of all this effort?
- What happens after death?

These are our spiritual limits—questions we edge closer to, but never fully resolve. And just like math, we must learn to live near the boundary, even if we don't cross it.

That's the art of existential courage.

Redefining the Function

Sometimes, to make a function continuous, we redefine it at the limit.

In life too, you can choose to redefine who you are—based not on where you came from, but on where you're going.

It's not cheating the curve. It's accepting that you were always bigger than your past slope.

Reflective Prompt: Find Your Limits

1. Where in your life do you feel yourself approaching a limit?
2. Is it emotional, relational, professional, spiritual?
3. How does your behavior change as you near it?
4. Can you observe it from both sides before reacting?

Closing Thought

"You are not defined by the value at the point of crisis. You are defined by how you approached it."

Limits are not dead-ends. They are turning points. They are where growth begins. Where change becomes irreversible. Where life becomes real.

So don't fear your limits. Approach them—slowly, wisely, and completely.

4: Continuity: Staying True to Yourself

Stability Through Change

If *differentiation* is the art of detecting change, and *limits* are the points where life turns—then *continuity* is the promise that, despite it all, something in you remains stable.

In calculus, a function is said to be continuous at a point if there's no break, no jump, no sudden hole in its graph. You can draw it without lifting your pen.

In life, continuity is your emotional coherence, your personal integrity, your ability to say:

"Yes, things are changing—but I'm still me."

It is not resistance to change. It is stability through change.

What Is Continuity?

Mathematically, for a function f(x) to be continuous at a point x=a, three conditions must be met:

1. f(a) is defined.
2. $\lim_{x \to a} f(x)$ exists.
3. $\lim_{x \to a} f(x) = f(a)$

In human terms:

1. You have a clear identity at the moment.
2. You have a stable approach toward that identity from all sides of experience.
3. You embody what you believe in—alignment between thought, intent, and action.

That is *continuity of the self.*

The Power of Being Unbroken

To be continuous doesn't mean never breaking down. It means—*even when life fractures*—your core stays traceable. Your story still makes sense. There is no sudden rewrite of who you are, just an evolution in expression.

For example:

- You lose a job, but not your work ethic.
- You end a relationship, but not your capacity for love.
- You face failure, but not the will to learn.

Continuity is you—unchanged at your foundation, even while your form adapts.

Emotional Continuity

We often mistake strong emotion for instability. But emotional continuity means being able to feel fully without losing the plot.

- You can cry and still be composed.
- You can be angry and still fair.
- You can be joyful without becoming careless.

Emotional continuity is not numbness. It's the ability to let emotion pass through you without distorting you.

Like a continuous function: There may be bends and twists, but never a break.

Discontinuity in the Human Curve

In math, discontinuity can come in several forms:

- Removable (a hole in the graph)
- Jump (a sudden leap)
- Infinite (asymptote)

In life:

- A removable discontinuity is a moment of regret—if resolved, the story flows again.
- A jump is a betrayal, a rupture, a sudden transformation.
- An infinite discontinuity is a breakdown—a space where the self collapses or explodes.

And yet, many people recover. Why?

Because continuity can be restored. Through healing. Reflection. Rewriting. Realignment.

You are never doomed by your discontinuities—unless you stop your function entirely.

Stability in Motion

We live in a world that glorifies change—pivoting careers, relocating cities, adapting to new trends. But without inner continuity, change becomes chaos.

Continuity means having:

- A set of values you live by
- A sense of self that doesn't dissolve under pressure

- A direction, even if your path is winding

It's your emotional home while everything else moves.

Real-Life Reflections

Think of people you admire—not for what they achieved, but for how they stayed grounded during success or loss.

- A leader who remained humble after fame.
- A teacher who never stopped learning.
- A friend who was always there, even when everything else changed.

These are continuous functions in human form. They make the graph of life traceable—for others and for themselves.

Continuity in Relationships

A healthy relationship is one where continuity is preserved through change:

- People grow, but they don't grow apart.
- They evolve, but the bond stays traceable.
- Arguments may bend the graph—but trust ensures no break.

Discontinuity in relationships happens not when people change, but when they stop relating.

Reflective Exercise: Your Continuity Check

1. What are the values you've held through most of your life?
2. When did you experience your biggest personal "jump"? How did you regain continuity?

3. What anchors you when everything else seems uncertain?

Write your answers like points on your personal curve. They are evidence that you've never really disappeared—only shifted slope.

Closing Reflection

"Continuity is not refusing to change. It is carrying your truth *through* change."

Your graph is long. It may have spikes and loops and asymptotes. But as long as it remains traceable, you remain whole.

So don't fear complexity. Fear only the moments when you can't recognize your own line.

Stay continuous. Not by being the same, but by staying *aligned to who you truly are.*

5: The First Ripple

Every Change Starts with Intent

Every great equation begins with a variable. Every great movement begins with an intent. And every ripple in life—visible or not—begins with something invisible: a decision to act.

In the Universe of Equations and Dynamics (UED), ripples are not random. They are born when energy shifts, when alignment occurs, when something deep within says: *"Now."*

And that *"now"* doesn't arise out of chance. It comes from intent—the most unmeasurable, yet most powerful force in life.

What Is the First Ripple?

In the physical world, a ripple starts when you disturb a surface. But in life, a ripple starts when your inner state breaks symmetry—when calm gives way to action.

- A thought becomes a plan.
- A desire becomes a step.
- A frustration becomes a change.
- A realization becomes a revolution.

That is your first ripple.

In mathematics, we often treat "change" as something already in motion. But in life, we must ask—*what caused the change to start?*

Intent is that zeroeth derivative. It's the pre-cursor. The shift *before* the shift.

The Power of Pre-Action

Intent is invisible. It doesn't show up in numbers. It can't be measured with tools.

But the entire trajectory of a function is decided by what you feed into it—initial conditions, limits, constraints. In life, those initial inputs come from your mental state, emotional charge, and clarity of will.

No ripple ever spreads without a center. And no center exists without intent.

Intent as Alignment

In UED, alignment between observer, field, and ripple triggers perception. In life, alignment between mind, emotion, and purpose triggers action.

When you intend something deeply:

- Your thoughts align with your feelings.
- Your words begin to reflect your truth.
- Your actions start creating consequence.

This is the birth of your personal energy dynamic—the origin of your function.

Ripples Begin Before They're Seen

Change doesn't start when people notice it. Change begins before—in the silent lab of your mind.

Think of the first time you:

- Decided to leave a toxic job
- Began learning something new
- Promised to never repeat a mistake
- Stood up for what you believed in

These moments didn't make the news. They didn't announce themselves. But they created ripples that changed your life—and others' lives too.

The universe doesn't need a loud bang. It listens to internal shifts.

Mathematical Analogy: The Seed of a Function

Before you graph a curve, before you solve a differential equation, you must define:

- Initial value
- Boundary condition
- Domain of interest

These are not decorations. They are everything. They decide the behavior of the system.

In human life, intent is that starting point. Everything else—emotion, consequence, perception—is the ripple that follows.

When Ripples Go Wrong

Sometimes, the intent is unclear. The emotion is noisy. The action is rushed.

This creates distorted ripples—misaligned actions, misunderstood decisions, unresolvable consequences.

But this too is natural.

Even in water, not every ripple is symmetrical. But it still originates from a center.

In UED, distorted ripples are not failures. They are just waiting to be re-aligned.

Reflection Exercise: Identify Your First Ripple

Think of something that changed you recently:

- Trace it back to the moment you decided.
- Go even further—what triggered that decision?
- Was it pain? Curiosity? A pattern? A voice?

That moment—however small—is your first ripple. Write it down. Honor it. That's where your calculus began.

Closing Thought

"You are not changed by what you do. You are changed the moment you decide to do it."

Intent is the whisper before the wave. The stillness before the shift. The prelude to all consequences.

So listen closely. Because often, the loudest ripple is the one no one else hears— except you.

6: The Rate of Emotion

How Feelings Evolve, Fast or Slow

Emotion is not a constant. It flows. It surges. It fades. It never stays the same. But more important than how we feel... is how fast we feel it.

In calculus, this is called a rate of change. In life, this is called emotional velocity.

Just as we study how fast position changes (velocity), or how fast temperature rises (rate of heat transfer), we must ask:

How fast does love grow? How quickly does anger escalate? How suddenly does fear paralyze you?

These aren't poetic questions. They are emotional derivatives—the calculus of our inner world.

Emotion as a Function

Every emotion is a function of time, experience, memory, and alignment.

You don't feel joy out of nowhere. You don't become depressed overnight. You don't heal all at once.

Each emotion follows a curve, sometimes exponential, sometimes linear, sometimes delayed. Some are sharp spikes; others are long waves.

- Grief lingers.
- Excitement peaks quickly.
- Trust builds slowly.
- Anger often spikes, then decays.

These are not flaws—they're signatures.

Derivative of Emotion = Awareness

In calculus, the derivative of a function tells us how rapidly that function changes. If emotions are functions, then your awareness is their derivative.

When you notice how fast something is rising within you, you're performing mental calculus:

- "I'm getting irritated."
- "This joy is wearing off."
- "This sadness is building too quickly."

That's your internal slope detector at work.

And the sooner you sense the slope, the better you can respond—before the emotion peaks or collapses.

Emotional Acceleration: When Slope Itself Changes

Sometimes, it's not the emotion itself that troubles us—but how fast it changes.

- You feel okay → suddenly overwhelmed.
- You love someone → suddenly resent them.
- You're calm → and then you're in tears.

This is not just rate of emotion—it's acceleration. In math, it's the second derivative.

In life, it's the surprise factor—when feelings evolve faster than your ability to understand or control them.

Emotional intelligence is not just knowing what you feel, but how fast you're changing—and why.

Flat Lines and Emotional Burnout

Just like in a graph, sometimes your emotional function goes flat:

- No joy.
- No sadness.
- Just numbness.

This is where the derivative is zero. No change. No movement. Often a sign of emotional exhaustion or suppressed feeling.

This state might seem stable—but it's dangerously quiet. Because if nothing changes, nothing heals.

Even a small ripple of feeling is a sign of life. Even a weak slope is better than none.

Slow Feelings Are Not Weak Feelings

In a world of instant reactions, slow emotions are often underrated.

- Love that builds over years is stronger than love at first sight.
- Trust that grows over time is harder to break.
- Forgiveness that takes its time is deeper than forced closure.

Not every emotion needs to race. Some need to breathe.

In calculus, not all functions are sharp curves—many are smooth transitions. That's how you build emotional resilience.

Real-World Reflections

Let's say you're teaching, leading, parenting, or mentoring someone. If you only see the final emotion, you might misread the person. But if you track how their feelings evolved, you'll see patterns:

- Were they always anxious, or did it escalate over time?
- Did their resistance build gradually or arrive suddenly?
- Is their current joy sustainable or momentary?

This emotional slope analysis helps us respond, not react.

Reflection Questions: What's Your Emotional Derivative?

1. What emotion do you feel most often, and how fast does it arise?
2. Do you tend to react quickly or slowly to emotional triggers?
3. What's one emotion you've learned to slow down—and how?

Write these answers down. Not to judge yourself. But to understand your function.

Closing Reflection

"Emotions don't lie. But their slope tells you more than their value."

Feelings are natural. But without awareness of their rate, they can mislead us—rushing decisions, derailing focus, or numbing experience.

So don't just ask, *"How do I feel?"* Ask, *"How fast am I feeling it?"*

Because the art of emotional calculus...is knowing when to pause, when to breathe, and when to let the ripple slow.

7: Velocity of Thought

Mental Sharpness as a Derivative

Your brain is not just a container of thoughts. It's a dynamic system, constantly evolving, adjusting, and responding—like a living function.

And when we talk about a *sharp mind*, what we're really describing is this:

The rate at which your thoughts evolve. The speed with which you process, pivot, and respond.

This is what we call the velocity of thought.

In calculus, velocity is simply the derivative of position—how fast something is changing at any given moment. In human terms, the velocity of thought is your mind's ability to:

- Recognize a problem.
- Connect variables.
- Predict the consequence.
- Adapt—before the rest even blinks.

It's not just about *what* you think. It's about *how quickly* your thinking adapts to new input.

Mental Velocity ≠ Intelligence

Let's clarify this up front.

Velocity of thought is not IQ. It's not about storing formulas or solving puzzles faster than others.

It's about agility. A person with mental velocity may not know all the answers—but they know how to ask the right questions. They respond to patterns. They *notice* shifts in context. They think in motion.

Intelligence is the database. Mental velocity is the processor.

Both are important. But in high-stakes environments—classrooms, surgeries, boardrooms, or battlefields—it's velocity that often makes the difference.

Thought as a Tangent Line

Imagine your mind like a curve.

Each idea is a point. Each new idea forms the slope of your thought process.

- A steep slope means sharp mental movement: quick realization, witty response, rapid learning.
- A flat slope suggests slow mental engagement, hesitation, or mental fatigue.

In classroom terms, this is the difference between a student who connects topics instantly versus one who needs repeated exposure.

But this slope can change over time—it can be trained.

What Affects Your Thought Velocity?

Several factors control the *derivative of your mind*:

1. **Interest and Alignment:** When you care deeply, your mind accelerates. Alignment boosts clarity and cuts hesitation.
2. **Sleep and Energy:** A tired brain is like a flat graph—no peaks, no progress.
3. **Mental Load:** Too many variables? The function becomes chaotic. Your slope can't stabilize.
4. **Confidence:** Fear slows thought. Confidence sharpens it.
5. **Environment:** A stimulating environment triggers faster pattern recognition—like being in a room full of thinkers.
6. **Practice:** Just as athletes train for speed, thinkers can train for velocity—through deliberate, repeated engagement with complexity.

Examples of Mental Velocity in Action

- A comedian delivers a spontaneous comeback—high slope.
- A chess master sees five moves ahead in a glance—instant derivative.
- A teacher adjusts lesson flow by reading the class energy—real-time thinking.
- A scientist sees a pattern no one else connects—curvature detection.

This is not memorized knowledge. It's motion knowledge.

What Slows the Mind?

- **Overthinking** – looping inside thought spirals instead of moving ahead.
- **Fear of failure** – second-guessing every mental step.
- **Disinterest** – lack of alignment brings inertia.
- **Excessive stimulation** – too much input can overload the function, leaving no clean derivative.

Think of it like noise in data—too much clutter hides the true slope.

UED Perspective: Thought as Ripple Derivative

In the UED model, ripple behavior reflects internal field alignment. Similarly, thought velocity is the ripple slope of the mind's energy field.

When the mind is aligned with intent and clarity:

- It forms high-quality ripples.
- The derivative (thought speed) is meaningful.
- Consequences are predicted better.

When misaligned:

- The field is noisy.
- Derivatives fluctuate—thoughts become erratic or dull.

Thus, mental sharpness is a field-readiness phenomenon—not a talent you're born with, but a state you align into.

Thought Exercise: Observe Your Own Mental Slope

- When was the last time you solved a problem

instantly?
- What were the conditions? Were you aligned, rested, focused?
- When was your mind slow? What burdened it?

Try keeping a *mental velocity journal* for a few days:

- Rate your thought speed (1–10) daily.
- Note what influenced it.
- Watch the patterns emerge.

You'll begin to see how emotional, physical, and contextual factors shape your internal derivative.

Closing Reflection

"A sharp mind doesn't know everything. It knows when to move—and when to pause."

Mental sharpness is not the peak of intelligence. It's the slope—the rate of rise of awareness, insight, response.

It's how fast a ripple resolves. How well a function adjusts. How smoothly the mind flows from *knowing* to *acting*.

So sharpen not just what you think—but how fast you ripple forward.

8: Slopes and Situations

Reading the Direction of Your Life

You don't need to be a mathematician to know when your life is going uphill—or sliding down. We all feel the gradient. We sense when things are accelerating, when progress is peaking, or when the weight of circumstances pulls us backward.

In calculus, this gradient is called the slope. Technically, it's the derivative of a function—how fast the function is increasing or decreasing at a specific point.

But in life?

The slope is the signal hidden in every situation. It's the direction you're moving in—not the value you're holding.

Two people can be at the same spot—same job, same income, same routine—but if one is on an upward slope and the other on a downward one, their lives feel completely different.

One feels progress. The other, decline.

Life Is Not a Straight Line

We often expect life to move linearly:

- Study → Graduate → Get job → Grow → Retire.

But real life doesn't obey such simplicity.

Instead, it curves. It climbs. It stalls. It drops. It takes turns—both expected and sudden.

And what matters most isn't where you are—but what direction your slope is pointing.

This is how we should read our situations:

- Are you growing or stagnating?
- Is this challenge pushing you toward growth or breakdown?
- Are you plateauing before a breakthrough—or about to fall?

These are slope questions—not position questions.

Positive Slope ≠ Easy Life

When life has a positive slope, things aren't always smooth.

You might be exhausted, facing tough decisions, dealing with chaos. But deep inside, there's a sense of movement. You know this discomfort is transitional—not terminal.

That's growth.

A positive slope means:

- You're learning.
- You're realigning.
- You're not stuck.

Even if today's value feels low, the slope tells you: *You're going somewhere.*

Negative Slope ≠ Rock Bottom

Similarly, a negative slope doesn't always mean failure.

It could mean:

- You're shedding what no longer serves.
- You're passing through discomfort toward clarity.
- You're being tested for alignment.

In physics, downward slopes accelerate objects. In life, they accelerate realization.

Sometimes, we need that decline to understand what we truly want.

What matters is how long you stay there, and whether you're aware you're descending.

Rock bottom isn't the end. A flat slope with no realization is worse.

When the Slope Is Zero

What happens when the derivative is zero?

That means the function is not increasing or decreasing. It's still. It's paused.

This is the plateau phase in life:

- Nothing major changes.
- No peaks. No dips.
- You're just... here.

Many people panic here. But not all stillness is bad.

In fact, this zero-slope moment is crucial for:

- Reflection
- Integration of past experiences
- Preparing for the next incline or decline

Just as in mathematics, a zero derivative could mean a turning point—either a maximum (you're about to descend) or a minimum (you're about to rise).

So ask yourself:

"Am I in a valley, waiting to rise?" Or, "Am I on a peak, avoiding the fall?"

Inflection Points: When the Curve Changes Its Mind

In calculus, an inflection point is where the slope changes its nature—from increasing to decreasing, or vice versa.

In life, these are the moments that redefine you:

- A breakup that leads to self-growth
- A failure that sparks reinvention
- A conversation that shifts your worldview
- A crisis that triggers clarity

You can't always predict them. But you can sense them—if you're aware of your slope.

Life doesn't change at extremes. It changes at inflections.

Reading Situations Like a Graph

Treat every situation as a point on a graph:

- What's the current value? (Your position)
- What's the slope? (Your direction)
- What's the curvature? (Your acceleration or fatigue)
- What's the context? (Where this point lies in your journey)

This mental habit will help you make non-emotional, high-awareness decisions.

When others ask, "How are you doing?" Don't answer with where you *are*. Answer with where you're *headed*.

UED Perspective: Ripple Slope as Life Direction

In UED, every ripple's slope carries energy—either in harmony or misalignment.

When ripples are outward, clear, and stable, their slope pushes energy forward. This is constructive slope—you move with the universe.

But when ripples are chaotic or collapse inward, their slope becomes destructive—you consume energy but go nowhere.

So in UED terms:

"Slope is not just movement. It is *direction of consequence*."

That's what life is.

Not your status. But your ripple slope through time.

Reflective Prompts

1. What area of your life currently has the steepest positive slope?
2. Where do you feel a negative slope creeping in?
3. Are you in a plateau, and what clarity is it offering?
4. Have you recently passed an inflection point? What changed?

Write your answers not to fix the slope—but to *observe it*.

Closing Thought

"You don't have to be at the top. You just need to know which way you're moving."

We are not defined by snapshots. We are defined by the flow—the continuous change of who we are becoming.

So next time you feel lost, don't ask, *"Where am I?"*

Ask, "What is the slope beneath my feet?" Because that's where the story of your life is quietly being written.

Interlude: The Function Comes Alive

We began this journey not with equations, but with experience.

Part I was never about mathematics in the conventional sense. It was about decoding life's rhythm—the calculus beneath consciousness.

Every moment we live, we generate ripples. Every thought, emotion, and decision carries slope, limit, and change.

Through these first eight chapters, we've seen:

- That we are not constants, but living functions—shaped by time and consequence.
- That differentiation isn't just math—it's awareness. The ability to measure our "now".
- That limits aren't constraints, but invitations to evolve. Life begins at thresholds.
- That continuity isn't perfection—it's emotional and intellectual integrity through transitions.
- That intent births the first ripple. Without it, the graph of life is undefined.
- That emotion has velocity—some feelings flash like

lightning, others evolve like glaciers.
- That thoughts, like particles in motion, have sharpness when aligned with direction.
- And finally, that life situations can be read like slopes—revealing not where we are, but where we're going.

This is not abstract metaphor. This is real-time life diagnostics.

You may not write these in notebooks or plot them on graphs, but your body, mind, and spirit compute this calculus every single day.

And the most beautiful part? None of this is reserved for the elite. No degree required.

Every human, knowingly or not, is a living equation. Every action is an evaluation. Every memory is an integral. Every realization is a derivative.

From Foundations to Form

Part I gave us the language of living change.

Not to impress, but to express—to help us observe ourselves without judgment, and others without haste.

Because once we understand the mechanics of change, we stop fearing it.

We start reading life, not as chaos or coincidence, but as consequence—a field of responses rippling with logic, emotion, and energy.

Now, as we move to Part II, the journey deepens.

We will begin integrating—not just observing what changes, but what remains. We'll explore how life accumulates meaning, memory, values, and patterns. We'll understand how we get stuck, where we stretch, and what it really means to grow.

Because after all, derivatives show us who we are becoming...but integrals show us what we've become.

Let's turn the page.

The story isn't just unfolding. It's being resolved—one ripple at a time.

Part II – The Integration Journey

9: Integration: Remembering the Story

Accumulating Experience, Not Just Solving Problems

In school, we're taught that integration is the reverse of differentiation. Mathematically, it means accumulating change over time.

But in life?

Integration is the act of remembering. It's not about solving problems—it's about honoring the process that shaped you.

Every moment you live, every choice you make, adds to something larger. Not every piece makes sense in isolation. But when you integrate them, they begin to tell a story—your story.

The Accumulated Self

Differentiation shows us how we change. Integration shows us what that change added up to.

It gathers all the little moments:

- The times you failed and got back up
- The thoughts that came and went
- The emotions that surged and faded
- The lessons you didn't know you were learning

Integration doesn't ask, "Did you succeed?" It asks, "What did it leave in you?"

Just like in calculus, we integrate with respect to time. In life, this means we learn not just from what happened, but when it happened, and how it was processed.

The Integral of You

Think of yourself as a function. Every second you live is a tiny slice of your curve. Taken individually, these slices are incomplete.

But when you integrate over time, you don't just get a number—you get a person.

- Your beliefs
- Your habits
- Your personality
- Your worldview

These aren't products of isolated events. They are accumulations.

You're not just the sum of your past. You are the integral of your lived experience—with a constant of uniqueness that no one else shares.

From Problem-Solving to Pattern-Seeking

Traditional education taught us to solve problems, get answers, move on. But life doesn't work like that.

In life:

- Problems repeat in disguise
- Solutions evolve with context
- Some questions don't have answers—just insights

- And some chapters are open-ended

Integration shifts your focus from solving to observing patterns—to noticing what keeps returning and why.

It helps you see:

- Why you're drawn to certain people
- Why some fears linger
- Why some joys feel eternal
- Why some choices repeat until you learn

You stop chasing solutions. You start building self-awareness.

Emotional Integration: Memory With Meaning

There's a difference between remembering and integrating an experience.

You can remember a heartbreak. But integrating it means understanding:

- What it taught you about boundaries
- How it revealed your capacity to heal
- Why it broke you—and how it rebuilt you

In UED terms, emotion is energy. And unintegrated emotion is misaligned ripple—a wave that hasn't been resolved yet.

Only when we integrate emotion with reflection does the field become stable again.

So many people carry trauma like unsolved equations. Integration doesn't erase them. It gives them meaning, so they stop interrupting your flow.

Integration Is Nonlinear

Just like in calculus, integration isn't always clean.

Sometimes:

- The limits aren't defined
- The function has discontinuities
- You need substitution, parts, or even partial fractions

Life too is like that:

- Some lessons take years to understand
- Some people are hard to forgive
- Some stories are fragmented, yet powerful

That's why integration isn't a one-time act. It's a life-long process.

You revisit old stories again and again—each time with more awareness, more compassion, more insight.

And every time, you reclaim a part of yourself.

The Constant of Life

In math, every indefinite integral includes a + C—a constant of integration. It accounts for what you cannot see directly but must include for completeness.

In life, that constant is your core self.

No matter what happens—

- Your values
- Your vision
- Your inner compass

—remain as a stable reference.

You change. You adapt. You evolve. But that constant is you, unshakable and timeless.

It's not the part you created. It's the part you remembered.

Reflective Prompts

1. What experiences have you lived through but never fully integrated?
2. Which past events still feel unresolved—not emotionally, but philosophically?
3. What patterns keep emerging in your life?
4. What is your personal "+C"—the constant part of you?

Closing Thought

"You are not what happened to you. You are what you've made of it."

To integrate is to reclaim. To connect the dots backward so you can step forward with strength.

Mathematics may define integration as an area under the curve. But life? Integration is the depth behind your story. It's the quiet meaning beneath the noise. It's the ripple that didn't fade—but found harmony.

10: Area Under the Curve

Every Moment Counts, Even the Small Ones

In calculus, the *area under the curve* tells us something profound:

Small changes, when added up, become something big.

A single point may seem insignificant. A short-lived emotion may pass without notice. A moment of clarity may vanish before it's acted upon.

But over time?

These moments accumulate. They form the curve of your life. And the area beneath it tells the story of what you lived, felt, learned, and became.

What Does Area Mean in Life?

In mathematics, the area under a curve represents total accumulation—of distance, energy, effort, or value—depending on the context.

In life, it represents everything that contributed, no matter how small.

- Every smile you gave
- Every fear you faced
- Every morning you chose to try again

- Every chapter you closed—voluntarily or not

Not all of them were grand. Not all went viral. But they added up—and still do.

Life is not just the curve you display. It's the area you carry inside.

Small Moments, Lasting Impressions

You may not remember:

- What you wore on your 17th birthday
- What your teacher said on a Tuesday in August
- What you were thinking while waiting in line five years ago

But those moments weren't empty. They were adding up, shaping the field under your existence.

Think of a cup slowly being filled. You don't hear each drop. But one day—it overflows.

That's what the area under the curve is: A silent overflow of invisible effort, unnoticed learning, and quiet growth.

Microefforts, Macro Results

We often overestimate the impact of grand efforts and underestimate the value of consistent, small ones.

- Writing 500 words a day doesn't look impressive—until you have a book.
- Exercising for 15 minutes a day feels trivial—until your body transforms.
- Practicing kindness in everyday life doesn't get headlines—but it builds character.

The curve doesn't care how steep you rise. It cares how long you sustained it.

And the area under it remembers everything—even the effort no one saw.

The Invisible Curriculum

Life teaches you in the margins.

- The job you didn't get.
- The apology you gave but never received.
- The idea you let go because it wasn't time yet.
- The Sunday evening you spent alone, reflecting.

These events don't get grades. They're not part of any formal syllabus. But they're part of your emotional calculus.

They shape your empathy, your resilience, your timing, your courage. They add area to your curve, layer by layer.

The Area of Relationships

A relationship is not defined by one event—good or bad. It's defined by the accumulated area of:

- Support
- Conversation
- Misunderstandings
- Forgiveness
- Patience
- Shared silence

If the area under your relationship curve is wide and full, one sharp dip won't destroy it. But if there's barely any area—just sporadic peaks and long gaps—the whole thing becomes fragile.

Emotional investment is cumulative, not event-based.

That's why trust takes time. That's why love grows quietly. That's why some people leave—not because of one moment, but because the area wasn't enough.

The Quiet Triumphs

You don't always know you're growing when you're growing. Some of your greatest chapters may feel uneventful while living them.

Like:

- Choosing to stay calm
- Learning to say no
- Letting someone else speak
- Trying again, just once more

Those may not spike your emotional graph. But they add volume to your soul.

They contribute to your personal integral—the story no one else can write.

Reflection Prompts

1. What small effort have you made consistently, that you never gave yourself credit for?
2. Which relationships in your life have large areas under the curve, and why?
3. How would your life look if you valued small moments as much as big ones?

Closing Thought

"The area under your life's curve is your lived legacy." It holds not just what you did, but how often, how deeply, and how silently.

It holds:

- The chapters you never published
- The ideas you nurtured
- The people you helped without a camera
- The love you offered without a name

Because in the end, the calculus of life won't be measured by the steepest peaks or the lowest valleys.

It will be measured by the total area of your intention, action, and presence— and how much of it you chose to fill with meaning.

11: The Constant of Integration

Honoring the Unknown

In calculus, every indefinite integral ends with a curious detail: A "+ C", the constant of integration.

It may seem like a footnote—a mere mathematical formality. But this single symbol hides a profound truth about life, self, and the unknown.

Because no matter how well you understand the curve...there's always something you didn't see—yet must acknowledge.

What Is the Constant?

In mathematical terms, the constant of integration exists because:

- Integration reverses differentiation.
- But differentiation loses constants—a flat horizontal shift vanishes in the derivative.

So, when we integrate, we must reintroduce the unknown lost in the process.

We say: "I don't know what was there before. So I must add something for completeness."

This is not weakness. It's intellectual humility.

In Life: The Missing Chapter

Apply this to life—and it becomes even deeper.

You may analyze yourself endlessly:

- Your thoughts
- Your habits
- Your memories
- Your turning points

But still, something remains unexplainable.

Why did you feel what you felt? Why did a moment hit you so hard? Why didn't a failure break you—or a success fulfill you?

There's no data for that. No formula.

That's your "+ C." Your constant of integration. The unknowable part of you that still matters.

The Constant Is Not a Flaw

In modern society, we're trained to explain everything.

We want closure. We want clarity. We want the full equation.

But life isn't a neat graph. It has blank spots, forgotten paths, invisible variables.

The constant isn't a flaw in logic—It's a reminder of the sacred unknown.

It represents:

- Your intuition
- Your faith
- Your past lives, perhaps

- Your subconscious drives
- Or maybe just... the mystery that keeps you human

You don't need to explain everything to live wisely. You need to honor what you can't explain.

Science and the "+C"

Even the greatest theories have constants.

- Planck's constant
- The gravitational constant
- The cosmological constant

These aren't just numbers. They are placeholders for what nature wouldn't reveal directly.

Even in science, constants emerge where understanding reaches a limit.

We measure the consequences. We infer the presence. We honor the gap with a symbol.

The constant doesn't lie. It whispers: "There's more to this than you know."

The Soul's Constant

In the UED framework, ripples align to consequences. But even then, there's a part of the field that remains untriggered, untouched—Not because it lacks energy, but because it hasn't met the right question yet.

This is the constant of your soul.

It may not appear in your visible actions, or even in your mental thoughts. But it's there.

- It's the reason two people with identical pasts live

different lives.

- It's why someone breaks and someone else bends.
- It's the intangible core that mathematics can only nod to—but never define.

Living With Your Constant

To live fully is not to solve everything. It is to accept that some part of you will always remain undefined—and yet deeply real.

That's not ignorance. That's wisdom in motion.

The constant of integration is where:

- Faith meets logic
- Art meets science
- Mystery meets method

It's the stillness within the ripple.

Reflection Prompts

1. What parts of yourself do you sense, but cannot explain?
2. Are you trying to define everything in your life before moving forward?
3. Can you love the mystery as much as the clarity?

Closing Thought

"The constant is not what you know. It is what makes you whole."

In calculus, it completes the equation. In life, it completes you.

Don't be afraid of what you don't understand. Let it breathe beside your understanding.

Because wisdom is not just measured by what you calculate.

It is defined by how deeply you respect the uncalculated.

And the journey continues—with C intact.

12: From Cause to Consequence

The Real Meaning of ∫

Among all the symbols in mathematics, few carry as much silent gravity as the elegant curve of the integral sign:

$\int$

At first glance, it looks like an elongated 'S'. But in truth, it represents the sum of infinitely small effects— the accumulation of consequences born from an underlying cause.

And in that, it mirrors life.

Because everything we are today...everything we feel, know, regret, love, or believe...is an integration of past ripples—some chosen, some accidental, some inherited.

The Integration Symbol: A Story in a Stroke

The ∫ is not just about calculation. It is philosophy in a symbol.

It means:

"Take everything that happened over this interval, multiply the change by its magnitude, and accumulate it into one, final consequence."

In simpler terms:

- What force did you apply?

- Over what duration?
- With what intensity?
- And what resulted from all that?

This is not unlike life's deepest questions:

- What did you stand for?
- What did you do with your time?
- What was your impact?

The integral doesn't measure moments. It measures meaning—in accumulated form.

Cause vs. Consequence

In UED (Universal Energy Dynamics), the ripple is primary. Not force. Not motion. Not probability.

Everything begins with intent—with alignment—with a cause that sends a ripple.

But what's the point of a ripple unless it meets consequence?

A ripple alone is just disturbance. A ripple that resolves... becomes reality.

This is exactly what the integral does: It takes the continuous field of causes and collects the consequence.

Just as:

- Light doesn't exist until it is seen.
- Emotion doesn't matter until it is expressed.
- Knowledge isn't wisdom until it is lived.

An integral is the bridge from concept to consequence.

∫ is not Summation. It is Transformation.

Many confuse the integral as a mere sum. But it's more than adding values. It's weighing them, contextualizing them, merging them into continuity.

In life:

- A kind act done at the right moment weighs more than a dozen missed ones.
- A choice made under pressure may define your entire arc.
- One realization can rewrite decades of belief.

The integral understands this. It doesn't just count—it honors what mattered.

The ∫ is not about quantity. It's about consequence density.

The Integral as Reflection

Every night, when you lie down and stare at the ceiling...What are you really doing?

You're integrating.

You're asking:

- What did I do today?
- Did it matter?
- What does it mean for tomorrow?

This is not abstract.

This is daily integration of life.

You scan your actions, filter through emotion, and extract consequence.

That's ∫ in motion—not on a graph, but in your soul.

What You Integrate Defines You

A powerful truth lies hidden here:

You don't become what you experience. You become what you integrate.

Many people live through hardship, but only some grow from it. Many gain knowledge, but few absorb it as wisdom. Many hear truth, but only some allow it to ripple inward.

Your life isn't the raw data. It's the integrated result of how you handled it.

The Boundaries of Your Integral

An integral always comes with limits:

$\int f(x)\, dx$

That's life too.

You can't integrate forever. You don't live from minus infinity to infinity.

You live from:

- Childhood to maturity.
- Ignorance to realization.
- Fear to courage.

Each phase has boundaries. And in that bounded interval, your experiences multiply with who you were in that moment.

And what emerges...is you—not as a value, but as a consequence field.

From Ripple to Resolution

In physics, integration gives displacement from velocity, energy from force, and motion from acceleration.

In UED, integration resolves ripple responses into manifested form.

In life, integration gives:

- Wisdom from experience
- Growth from pain
- Identity from interaction
- And stories from silence

It's not the equation that defines your life.

It's what you did with the variables, and how deeply you carried the result.

Reflection Prompts

1. What experiences have you lived—but not yet integrated into learning?
2. Are you focusing more on cause, or on understanding consequence?
3. Can you see a phase of your life as an interval—from intent to outcome?

Closing Thought

∫ is not a symbol of calculation. It is the signature of consequence.

It reminds you that:

- You were not just there—you left an effect.
- You didn't just feel—you transformed.
- You weren't just changing—you were becoming.

You were not the cause alone. You were the consequence, written by ripples. And in that sense...You are your own integral.

13: Substitution: Reinventing Ourselves

Change of Variables in Identity

In calculus, the method of substitution is a clever trick: When an expression becomes too complex to resolve directly, we don't fight it. We change the variable. We shift the frame. We temporarily reimagine the problem—so it becomes solvable.

That's not cheating. That's intelligence in action.

And so it is with life.

When the problem becomes too heavy, too confusing, or too deeply entangled with our existing identity, we must allow ourselves the freedom to substitute—to reinvent.

u = g(x): A New Perspective

In integration by substitution, we say:

Let $u = g(x)$

Why? Because working directly with x is complicated.

So we:

- Shift to a simpler expression (u)
- Solve in this new variable

- And then return to x—having resolved the chaos

This is exactly what we do when we:

- Take a break from our role to reflect
- Change environment to regain clarity
- Step into a new identity to grow beyond limitations

Substitution is not rejection. It is reorientation.

You're not denying who you are—you're simply rewriting the problem in a solvable way.

Reinvention Is a Natural Process

We are not born fully defined. We change roles, names, professions, beliefs, and sometimes entire personalities.

And that's not hypocrisy. That's growth.

As long as the core continuity remains—the curve of life is not broken. It is simply re-expressed in new variables.

- A teacher becomes a mentor.
- A soldier becomes a monk.
- A scientist becomes a poet.

In each case, the soul didn't change. Only the variable of expression did.

And with that change, previously impossible integrals of life—suddenly become solvable.

Who Are You, Really?

Ask yourself:

- Are you the career you chose?
- The language you speak?

- The religion you follow?
- The body you occupy?

No.

These are variables—not constants. And when one of them becomes too rigid or too painful, it may be time to substitute.

Let:

u = the wiser you

u = the more peaceful you

u = the courageous version you've been postponing

And see what integral becomes solvable then.

Substitution ≠ Escape

Substitution is often misunderstood as escape. But in calculus, we substitute not to avoid complexity—but to resolve it.

Similarly, reinventing yourself is not about running away from who you were— It's about transforming your approach so you can finally meet the consequence.

Because life will not stop challenging you. It will just keep changing form.

And you must learn to respond by changing your variables, not by erasing your function.

UED: Variable Alignment Across Fields

In UED terms, substitution is ripple redirection.

The ripple is still you—but its alignment across different fields (mental, emotional, professional) may require different expressions of the same intent.

Just as light bends when it moves through different media, so do you—when passing through the various layers of life.

What matters is not the variable.

What matters is:

Does the ripple continue?

If it does, identity remains preserved, even if the variable has changed.

Substitution Moments in Life

Here are real-life substitutions:

- After a heartbreak, you don't abandon love—you substitute expectation for understanding.
- After career failure, you don't give up—you substitute ambition for curiosity.
- When aging, you don't lose relevance—you substitute performance for wisdom.

Each of these is a change of variable —allowing the integral of your life to proceed.

You are not fake when you change. You are evolving.

Reflection Prompts

1. Are you currently struggling with a life expression that feels too complex?
2. What identity shift could simplify or clarify your next step?
3. Can you redefine yourself without denying your core?

Closing Thought

"When life gets too complicated, change the variable—not the vision."

You are not abandoning who you are. You are simply rewriting your function in a way that allows resolution.

Because sometimes, the problem of life doesn't need brute force.

It needs a better substitution.

And with that shift, a new version of you begins to integrate.

14: Chain Reactions

When One Life Depends on Another

In calculus, there's a beautiful concept called the chain rule. It tells us that the rate at which one variable changes depends not just on itself—but also on another variable it's linked to.

Mathematically, it's expressed as:

$$d/dx\,[f(g(x))] = f'(g(x)) \times g'(x)$$

One function nested inside another. One consequence dependent on another's cause.

And isn't that just how life works?

Your story is not yours alone. It is interwoven—layered within the actions, emotions, and intents of others.

We Are Nested Functions

You didn't become who you are in isolation.

- Your childhood dreams were shaped by your parents' fears.
- Your beliefs by the culture that enveloped you.
- Your triggers by someone else's unresolved trauma.
- Your courage by someone else's sacrifice.

You are a function of another function. And many others are a function of you.

This is the chain rule of life.

We ripple because someone else did.

The Mathematics of Mutual Consequence

The chain rule teaches us that:

Even a tiny ripple at the inner level ($g'(x)$) can amplify or dampen the overall change ($f'(g(x))$) at the outer level.

In life:

- A kind word from a teacher may change the student's entire trajectory.
- A moment of betrayal may spiral into someone else's lifetime of mistrust.
- A father's silence might become his child's unresolved fear of expression.
- A friend's belief in you might trigger your long-awaited transformation.

Your *derivative*—your impact—is amplified through others.

That's why you matter more than you think.

UED and Ripple Linkages

Under Universal Energy Dynamics (UED), no ripple is isolated.

Every ripple that emerges from one source becomes a disturbance—an influence—in the neighboring field. This leads to a chain of realignments.

In this light:

- A personal act of clarity can purify a group.
- A social misalignment can spread as collective unrest.
- A teacher's clarity may ripple through generations of students.

This is not poetic exaggeration. This is the physics of connection.

Chain reactions are not metaphors. They are how alignment spreads.

Chain Reactions in Relationships

Think of your life like nested functions:

- Your partner's mood may affect your decisions.
- Your child's struggle might redefine your priorities.
- Your boss's pressure might manifest in your own tone toward others.
- Your friend's honesty may unlock your buried truth.

We often act as if we are self-contained units. But reality shows we're deeply entangled.

Your change is someone else's derivative. And theirs is yours.

Responsibility in the Chain

The chain rule demands responsibility. It means:

Before you judge the outcome, understand the input.

Before you criticize someone's reaction, ask what triggered it.

This doesn't mean we excuse harmful actions. But we trace the origin—not to forgive blindly, but to understand deeply.

Because healing doesn't just mean correcting the outer function. Sometimes, it means going back to the original g(x) that shaped it all.

When You're the Inner Function

You may not always be the one facing the world.

Sometimes:

- You're the hidden g(x)—the quiet cause.
- You're the voice behind someone else's success.
- You're the invisible mentor whose derivative shaped a life.

Don't underestimate this role.

You may never be famous. But you may be formative.

Healing Broken Chains

What happens when the chain breaks?

When someone in the loop causes trauma instead of growth?

Then, the new function—you—may need to take charge:

- Redefine your derivative.
- Set new boundary conditions.
- Begin a chain reaction of healing instead of pain.

You don't always choose your place in the chain. But you do choose whether to continue it... or redirect it.

Reflection Prompts

1. Whose chain reaction are you part of—right now?
2. Are you amplifying healing or carrying forward pain?
3. Who has been the hidden g(x) in your life?

Closing Thought

No one lives a derivative-free life.

You are someone's consequence. And you are someone's cause.

The moment you understand this, you don't just move with life.

You start moving with intention—aware that every shift in you might cascade into others.

Because in this interconnected equation of existence...

The chain never breaks. It just passes on.

Make sure what you pass is worth integrating.

15: Integration by Parts

Giving, Losing, Growing Through Others

In calculus, integration by parts is a method that tells us a profound truth:

When two functions intertwine, their integration is not just the sum of their individual paths. It's an intricate exchange—one gives while the other evolves, and both change in the process.

Mathematically, it's written as:

$$\int u \cdot dv = uv - \int v \cdot du$$

But beneath this equation lies a deeper life principle: You don't grow alone. You grow through others.

Life as a Shared Calculation

In the integration by parts formula:

- u is the part we hold on to—the stable identity, the will, the presence we offer.
- dv is the energy, the effort, the change that's happening—what moves around us.

As we proceed:

- u is differentiated → du: we lose a bit of ourselves, we transform.
- dv is integrated → v: someone or something grows through what we gave.

In essence:

- We give part of ourselves (du)
- And in doing so, the other side (v) evolves
- Then they return the favor—and their growth influences our outcome

This is the dance of reciprocity, relationship, and growth through interaction.

uv – The Moments of Meaning

In the formula:

$$\int u \cdot dv = uv - \int v \cdot du$$

That first term uv—it's the shared moment, the connection.

It's the moment:

- A mentor empowers a student.
- A parent sacrifices so the child can rise.
- A leader supports a team through struggle.
- A friend listens at just the right time.

These are the points of impact—where two lives intersect, and both leave a mark on each other.

They are finite, specific, and powerful.

They don't last forever—but they shape everything that follows.

$\int v \cdot du$ – The Cost of Growth

Then comes the second term: $\int v \cdot du$

Here's where life tells us something uncomfortable:

Growth comes with loss. Contribution involves cost. Supporting others often changes us—and not always in the way we expect.

You may:

- Give wisdom, and feel emotionally drained.
- Teach with passion, and find your own energy dimmed.
- Guide someone else, and start to question your own direction.

But this isn't failure. It's the price of connection.

In the ripple universe of UED, whenever energy realigns in one field, it draws from another.

This isn't destruction—it's redistribution.

Just like in integration by parts, we trade roles. We give, and then we change.

Roles That Reverse

In life, the u and dv roles keep switching.

- Today you are the teacher; tomorrow, the learner.
- Today you support; tomorrow, you'll be supported.
- Today you're the one giving light; tomorrow, someone else will hold the lamp.

Integration by parts teaches us to honor both roles.

Because the story of your growth is incomplete without the people who helped you —and the people you helped, even when it hurt.

UED: Field Interactions in Real Life

In UED terms, this process is not abstract.

Every ripple that interacts with another field redistributes its structure.

One field becomes slightly dimmer, the other brighter.

But together, they form alignment—the condition required for resolution.

Thus, integration by parts is not merely a technique—it's a philosophy of coexistence.

No ripple grows in isolation. Every expansion is fed by another's contraction.

Reflection Prompts

1. Who in your life has been your u—the stabilizing force that sacrificed to help you grow?
2. When have you been someone's dv—the one moving forward with borrowed strength?
3. Can you accept the cost of giving, without resentment?

Closing Thought

Life is not a straight integral. It's integration by parts.

We live, we lose, we support, we suffer. And in doing so, we grow not only ourselves, but others.

You are not a hero because you gave. You are a hero because you understood what you lost—and still chose to give.

Because in the calculus of life, the function is never solved alone.

It is resolved together, One part at a time.

Interlude – The Integration Journey

"To live is to remember, realign, and rise—one part at a time."

Differentiation teaches us how fast we're changing. But integration reveals what we've become.

This part of the journey is not about the moment-to-moment motion—it's about what's been carried forward, what's accumulated, and what's been left unresolved.

In mathematics, integration finds the area under the curve—but in life, it helps us discover the meaning behind the motion. It tells us:

Every step we took mattered. Even the smallest increments—the tears, the silences, the forgotten mornings—are part of the whole. And nothing truly disappears; it integrates into us.

In these chapters, we've begun to explore how:

- Experiences build memory
- Uncertainties become constants
- Cause merges with consequence
- Identities evolve through substitution
- Lives link through chain reactions
- Growth occurs when we lose, give, and trust others to

hold part of our story

This is not just a mathematical analogy. It's a lived truth. We integrate people, moments, dreams, regrets—folding them into who we are today.

Some integrals are straightforward. Some require parts. Some demand substitutions. Some leave us with constants—mysteries we can never resolve fully.

But the act of integration itself is sacred.

It reminds us that life isn't just about what we *solve*. It's about what we absorb, what we transform, and what we become capable of holding.

So pause here. Reflect.

What have you integrated so far?

Not just from this book—but from your own life.

What areas beneath your curve remain unseen? What constants remain undefined? What consequences are still echoing?

Because in Part III, we will begin to explore not just how to change or integrate...but how to align.

And that is where true resolution begins.

Part III – Complexity and Alignment

16: Partial Fractions, Complete Life

Breaking Down Complexity to Solve It

In mathematics, when an expression becomes too complex to handle directly— we don't fight it. We break it down.

That's what partial fractions do.

They take a tangled expression and decompose it into simpler, digestible parts. From chaos to clarity—one piece at a time.

And isn't that exactly how we deal with life?

When things get overwhelming, the only way forward is to simplify the impossible.

The Art of Decomposition

Take a rational function like:

$$\frac{P(x)}{Q(x)}$$

Where Q(x) is a complex polynomial—dense, heavy, full of turning points. You could try integrating it directly... but it won't yield easily.

So we split it.

We rewrite it as a sum of simpler expressions, each one more approachable. Each one solvable.

That's not just algebra.

That's psychology. That's survival. That's how we live.

When Life Gets Too Complicated

Sometimes:

- You carry unresolved grief and responsibilities.
- You juggle ambition with self-doubt.
- You are both healer and the one in need of healing.
- Your day is a blend of progress, pause, and pain.

Trying to solve all of it at once is emotional suicide.

But break it down?

- This is the morning tension → rooted in fear of failure.
- This is the unresolved argument → still echoing under your confidence.
- This is creative fatigue → misread as laziness.
- This is joy → fleeting but real.

And suddenly, you're not broken. You're layered.

Just like a complex expression, you aren't unsolvable. You're just made of parts that need individual attention.

The UED Insight: Field Complexity = Misaligned Layers

In UED, what appears chaotic in a ripple field is often just multiple ripples unresolved.

Complex patterns emerge when overlapping causes are still unresolved through consequence.

But if you trace them backward, one by one—if you break down the wave function—you find:

- A core emotion,
- A triggering event,
- A misaligned intent,
- A latent memory.

Each part has its own wavelength. Its own story.

Once identified, realignment becomes possible—not as a one-shot cure, but as a progressive disentanglement.

Partial Fractions in Human Relationships

Try applying this to people:

You're frustrated with someone close. But instead of labeling it as "bad energy", try this:

- Is it their tone of voice (A)?
- Or your expectation they didn't meet (B)?
- Or unresolved issues from someone else projected here (C)?
- Or simply your current emotional bandwidth?

Each is a partial fraction of the whole interaction.

Solving the whole expression without decomposing it only leads to more friction.

Once you see the parts, forgiveness, clarity, and communication become easier.

Decomposition and Identity

Even identity is a complex expression:

You are not just "you." You are:

- The student who once failed
- The child who was loved or ignored

- The friend who listened
- The worker who survived
- The thinker who questions
- The dreamer who never gave up

These are your partial selves.

You can't fully integrate unless you acknowledge each part, give it space, hear it out.

Your wholeness is not in pretending you're one smooth curve—It's in owning that you're made of many expressions, each with its own variable and weight.

Reflection Prompts

1. What's currently overwhelming you? Can you break it down into "partial fractions"?
2. Are you misjudging complexity as failure, instead of seeing it as layered structure?
3. Which part of yourself have you been ignoring that might be holding the key to resolution?

Closing Thought

Complexity isn't a curse. It's an invitation to understand life in layers.

Just as we use partial fractions to simplify the unsolvable, we must learn to decompose our emotional and existential density.

Not to minimize it. But to approach it—piece by piece—with compassion and clarity.

Because: A complete life is not built from perfection. It is solved by understanding each part for what it truly is.

17: Discontinuities and Trauma

Healing the Broken Curves

In mathematics, a discontinuity occurs when a function suddenly breaks— a sudden jump, a hole, or a point where the curve simply does not connect.

It signals an interruption in flow, a failure to align cause and consequence smoothly.

And in life, we know these points intimately.

We call them:

- Trauma
- Loss
- Abandonment
- Heartbreak
- Failure
- Betrayal

These are the broken points in our personal graph—where experience didn't just shift, it snapped.

The Function Was Smooth… Until It Wasn't

Sometimes, everything seems continuous. Predictable. Manageable.

Then:

- A parent walks out.
- A loved one dies unexpectedly.
- A friend turns against you.
- A job you built your identity around ends.
- A single mistake collapses years of confidence.

These events create emotional discontinuities.

The mind tries to continue functioning, but the curve no longer flows. The derivative explodes. The limit doesn't exist.

We're left disoriented—trying to understand how we got here, and why the math of life broke down.

What Mathematics Teaches Us About Healing

In calculus, when facing a discontinuity, we don't throw away the entire function.

We do three things:

Identify the type of discontinuity:

- Is it a removable gap?
- A jump discontinuity?
- An infinite asymptote?

Approach it from both sides:

- What happens as we near the point from the left (past)?
- What happens as we approach from the right (future)?

Reconstruct or redefine the function at the point of breakage, even if it requires new terms, external limits, or a patch.

This is exactly what trauma healing demands.

UED and Discontinuities: Ripple Collapse

In UED, trauma can be seen as a collapsed ripple, where incoming consequence couldn't be processed by the field.

The energy input exceeded field capacity. Misalignment peaked. The consequence fragmented instead of resolving.

But that doesn't mean healing is impossible.

What's needed is not more force—but controlled alignment, intelligent waiting, and field recalibration.

The broken part of the ripple still exists—but it's locked, waiting for resonance.

Approaching Trauma: The Left and Right Limits of Experience

Think of your trauma as a point of discontinuity at x=a

- **Left limit**: What were the events, beliefs, and assumptions that led you into it?
- **Right limit**: What did you do after? How did you adapt, avoid, or compensate?

Often, we live in a loop of avoidance, never truly analyzing either side.

But healing comes only when:

The left and right limits begin to converge.

When we can finally say:

"I understand why it happened. And I see how I've changed because of it."

Then, we are no longer a broken curve—we are a piecewise function, gracefully redefined.

Trauma is Not a Bug in the System

Here's the radical truth:

Trauma is not a failure of life—it's a boundary condition.

It is where:

- Memory gets stuck
- Patterns get interrupted
- Consciousness asks to pause, and re-author the function

It's painful, yes. But also powerful.

Because when you reconstruct a discontinuity, you don't just heal—you reshape the entire curve of your life.

Emotional Continuity: The True Goal of Healing

Healing is not forgetting.

Healing is redefining the function so the curve can move again. Even if it never looks like it used to.

You are allowed to:

- Grieve what was lost
- Redefine who you are
- Set new rules for the future
- And still call yourself whole

Because life, like math, doesn't require perfection.

It only asks for consistency, and the courage to patch what broke.

Reflection Prompts

1. Where in your life do you feel a break—a moment that left your story fragmented?
2. Are you avoiding that point, or can you now approach it from both sides—like left and right limits?
3. If you could redefine your function at that point, what would it say?

Closing Thought

Every life has discontinuities. Not all can be removed. But all can be approached—with understanding, patience, and truth.

You are not broken. You are a reconstructed curve. And in that reconstruction lies your greatest beauty.

18: Inflection Points

Life's Defining Turning Moments

In calculus, an inflection point is where the curve changes its nature— where it stops bending one way and begins to bend the other.

The slope doesn't vanish. The function doesn't break. But something subtle, profound, and irreversible happens:

The concavity changes.

That's an inflection point.

And in life, these moments are everywhere—not always dramatic, but deeply directional.

The Shift Beneath the Surface

Mathematically, the first derivative tells us the direction (positive or negative), but the second derivative tells us the behavior of change itself—is the function accelerating or decelerating? Bending upward or downward?

The inflection point is where that behavior flips.

Likewise, in life, we often continue on the same visible path—same job, same house, same daily routine—but something inside bends differently.

- A moment of realization.
- A conversation that lingers.

- A decision made in silence.
- A truth that lands in your heart like a pebble in still water.

The outer slope remains—but the inner story changes curvature.

Detecting the Invisible Turn

Inflection points are not always like trauma (discontinuities), nor are they always marked by extreme events.

They may look like:

- Choosing growth over comfort.
- Saying yes to something uncertain.
- Saying no after years of pleasing others.
- Forgiving someone, not for them—but for you.
- Letting go of a goal that no longer aligns.
- Falling in love—not with someone, but with yourself again.

The curve doesn't stop. It continues. But it starts bending toward something else.

Inflection Points in UED: Realignment Triggered by Inner Phase Shift

In Universal Energy Dynamics (UED), an inflection point is when ripple alignment shifts due to a change in internal phase or amplitude, even if energy flow continues.

There is no external collapse or discontinuity—just a directional correction in how consequence is moving.

You're not destroyed. You're not reborn. You are reoriented.

It's the moment the soul whispers louder than the noise.

Why We Miss Our Inflection Points

Unlike a dramatic turning point or crisis, an inflection point often comes quietly.

- There's no announcement.
- No applause.
- No tragedy.

That's why many people miss it—until years later, when they look back and say:

"That was the moment everything began to change."

But if you're mindful... You can feel the curve shift under your feet.

You may not know where it's going, but you'll know:

You're no longer going where you were.

Embracing Your Curvature

The function of life is not linear. We all bend, arc, hesitate, lean forward, recoil, and stretch again.

The most graceful among us are not those who stay straight, but those who honor their inflection points.

They adjust their course—not out of fear, but out of awareness.

Reflection Prompts

1. Can you identify an inflection point in your past—when your life began bending a new way?
2. Are you currently approaching one? What

signs—emotional, relational, spiritual—are showing you a curve is coming?

3. What would happen if you stopped resisting the change and allowed your path to arc naturally?

Closing Thought

Inflection points don't shatter you. They reshape you. They don't stop your journey. They redirect your horizon.

Let the curve of your life change. Let it bend into alignment, not just ambition.

Because some of the most important turns in life happen not with a bang, but with a soft change in curvature— felt more than seen, understood only in reflection.

19: Maxima and Minima

Success, Failure, and the Second Derivative of Ego

In calculus, a maximum is the peak—a local or global high point. A minimum is the valley—a low point where the slope flattens before climbing again.

Both are critical points, where the first derivative becomes zero. But not all zero slopes mean success or failure. To know what kind of point we are at, we use the second derivative.

And in life, that second derivative... is often our ego.

The Stillness at the Turning Point

At the very peak of success—or the depth of failure—life often feels still.

- You've climbed the mountain. Now what?
- You've hit rock bottom. What's next?
- You're neither rising nor falling. You're flat.

This is the zero-slope moment. Externally, everything looks calm. Internally, it's the most sensitive point—because the next step determines the new direction.

A moment too long in ego, and the peak becomes a plateau. A step too soon in fear, and the fall becomes a collapse.

The Role of the Second Derivative: Curvature of Intent

In calculus, the second derivative tells us the nature of change:

- If it's positive, you're at a minimum—things are about to rise.
- If it's negative, you're at a maximum—you may soon fall.
- If it's zero, the curve may flatten—an inflection or deception.

In life, this curvature is not defined by numbers. It's shaped by intent, awareness, and ego.

The ego says:

"I've arrived." "I'm ruined." "I'm better than others." "I don't deserve this."

The ego flattens your awareness. It hides the true curve.

But your inner intelligence—your UED field—knows the change before it happens.

Success is not about the peak itself. It's about how you move after it.

Ego: The Invisible Second Derivative

You don't always realize your ego has taken over until:

- You stop listening.
- You stop learning.
- You ignore alignment.
- You confuse fame with fulfillment.
- You see failure as identity instead of instruction.

That's when the curve starts turning, long before results show it.

The ego doesn't make you rise or fall—it decides the shape of your path after the slope flattens.

In UED terms, ego is stored ripple momentum trying to maintain a form. If not resolved, it resists the natural flow toward equilibrium.

Success and Failure: Just Local Extrema?

Life has no single global maximum. There are many peaks, and each one teaches a different lesson.

Likewise, failure is not a final minimum—it's a pause before the next ascent.

So rather than label your positions, ask:

- Am I climbing because I'm aligned?
- Am I falling because I'm resisting correction?
- Am I flat because I'm meant to turn?

Maxima and minima are feedback points.

They don't define you. They redirect you.

Reflection Prompts

1. What was your last emotional maximum? Did ego take over afterward?
2. Are you in a minimum right now? Can you feel the second derivative beginning to rise?
3. Do you mistake flatness for stability? Or are you ignoring a pending turn?

Cosmic Perspective

Even planets have maxima and minima in their orbits. Even light curves near gravitational fields. Even galaxies spin through peaks of creation and troughs of dissolution.

Why should your life be any different?

The law is universal: What rises with force must align with consequence. What falls with grace may rise through resonance.

Closing Thought

Success isn't your peak. Failure isn't your fall. They are just critical points on your evolving graph. The second derivative—your humility, curiosity, and alignment—determines the next shape.

Be cautious at the top. Be courageous at the bottom. Be conscious at zero slope.

Because the curve is always alive. And you're more than a point—you are the function itself.

20: Asymptotes and Acceptance

What We Approach But Never Reach

In the world of calculus, an asymptote is a boundary. A line the curve gets closer to... ever closer...but never touches.

No matter how far it travels, how much it grows, bends, or evolves, the function will never intersect it.

In life, too, we chase many such asymptotic ideals:

- Absolute truth
- Perfect love
- Complete peace
- Total success
- Eternal happiness

We get close—sometimes breathtakingly close. But just when we think we've arrived, something subtle pulls it a step farther.

That's not failure. That's life's design.

The Idea of a Perfect Destination

We are told to chase goals, fulfill dreams, seek enlightenment, or find "the one." And so we build expectations like straight lines on a graph—believing one day we will reach them.

But these ideals often behave like horizontal or vertical asymptotes:

- Horizontal asymptotes: Lifelong desires we get closer to as time moves forward, but never truly possess—like peace, wisdom, or contentment.
- Vertical asymptotes: Explosive emotional states—like obsession, grief, or ecstasy—that feel overwhelming near a certain experience but can't be sustained.

Try to *cross* them, and you break. Learn to *approach* them, and you grow.

This is not giving up. It's acceptance.

Acceptance is the Art of Asymptotic Wisdom

To accept doesn't mean to stop trying. It means to stop believing that touching the limit is necessary for meaning.

In UED terms, the asymptote is a resonance boundary—the limit where the ripple of experience is still felt, but cannot resolve into consequence.

Why?

Because some experiences aren't meant to be "solved." They are meant to be respected.

You don't solve grief. You integrate it.

You don't reach infinity. You curve toward it.

The Spiritual Asymptotes

In every ancient tradition, there's a divine idea always just beyond reach:

- Moksha
- Nirvana
- Heaven
- God

Not unreachable because they are fake— but because they represent asymptotic truths.

They guide our curve, not as endpoints, but as directions to align with.

The more we evolve, the closer we feel to them— yet we realize they were never fixed points. They were the infinite extensions of our finite lives.

Ego Wants Intersections, Soul Understands Approaches

The ego says: "I must reach. I must possess. I must conquer." The soul whispers: "Approach is enough. The journey is sacred."

This is the difference between chasing meaning and living with meaning.

It's the same difference between a curve desperately trying to intersect an asymptote, and one that aligns with it forever, without the need to touch.

Reflection Prompts

1. What in your life feels like an asymptote—something you're forever approaching?
2. Have you mistaken a divine direction for a final destination?
3. Can you live peacefully, knowing some things are meant to stay slightly out of reach?

Closing Thought

Not everything in life is meant to be achieved. Some things are meant to be witnessed, pursued, and cherished—without possession.

Asymptotes teach us the limits of control. They humble our ambition, not to weaken us—but to realign us.

Because acceptance is not the end of effort—it's the beginning of grace.

The curve continues. The line remains. The two never touch...and yet, their relationship is eternal.

Interlude – Complexity and Alignment

"*When the curve bends, so does our understanding.*"

Up to this point, we've learned how to recognize change and remember experience. Now, we face the real terrain of the journey—complexity.

This part is where life stops playing fair, and equations don't always solve cleanly. This is where:

- The function doesn't behave as expected.
- The curve breaks suddenly.
- The direction flips without warning.
- And the destination drifts beyond reach.

In mathematics, such situations are handled with advanced tools—partial fractions, discontinuities, second derivatives, and asymptotes.

In life, we handle them with:

- Reflection
- Resilience
- Realignment

- And eventually—acceptance

These chapters are not for the faint-hearted. They explore the trauma we avoid, the turning points we deny, and the limits we pretend don't exist.

But understanding complexity doesn't mean complicating our minds.

It means learning to see the parts, without losing sight of the whole. It means knowing you can heal, even when the function looks undefined. It means recognizing that your story doesn't peak at success—it curves through it. And it means embracing the idea that some goals are asymptotic—meant to guide, not to own.

This is alignment—not perfection. This is intelligence—not control. This is the part where the calculus of life becomes art.

Welcome to Complexity and Alignment—where your curve begins to understand itself.

Part IV – The Universal Field

21: The Ripple Equation of Life

UED as the Backbone of Calculus

"Every change you observe is not the reality itself, but a ripple seeking alignment."

In classical calculus, we model change through limits, derivatives, and integrals. We define motion as variation in position over time. We define growth as the area under a curve. But beneath these elegant notations and formulas lies a question most never ask:

What is it that truly changes?

In Universal Energy Dynamics (UED), the answer is both radical and disarmingly simple:

Nothing *really* changes. The universe is still. What we perceive as motion, transformation, or life is the ripple consequence of interactions within the structured vacuum energy (SVE) field.

This chapter introduces the heart of UED as the invisible backbone of what we call calculus. It rewrites the foundational ideas of differentiation and integration not just as mathematical tools but as philosophical truths of observation and perception.

The Premise: Change Is a Consequence, Not a Cause

Modern science begins with change. Newton asked, "What causes acceleration?" Calculus answered with derivatives and limits. But UED steps behind that curtain.

It says: Change is not a first principle. It is a perceptual consequence.

In the ripple field view, everything exists in a state of energetic balance unless disturbed. Any force, movement, light, or sound is not an independent entity but a disturbance propagating through the structured vacuum. This disturbance takes the shape of ripples—not in water, but in the field that underlies matter and space.

What we differentiate in mathematics is the rate of perceived change. What we integrate is the accumulated impact of that perception.

Both are consequences of ripple dynamics.

The Living Function Revisited

Earlier, we explored the idea that a person is not a constant. We evolve, adapt, react. But from the UED perspective, the person isn't moving at all. The ripple pattern we identify as "Sandeep" or "Ravi" or "Meera" evolves in the universal field like a standing wave modulating with experience.

Your thoughts, memories, emotions, and actions are not particles darting through neurons. They are ripple alignments, forming meaningful patterns. These patterns, when unresolved or misaligned, appear as distortion, tension, or mental noise. When aligned, they feel like clarity, flow, and peace.

The calculus of life is then not about change *in you*, but about ripple responses *around you*.

UED Formulation of Derivatives: Observation as Displacement

Let's reinterpret the first derivative.

In classical terms:

$$f'(x) = \lim (h \to 0) [f(x+h) - f(x)] / h$$

In UED:

f'(x) represents the first observable ripple deviation in the field when a unit perturbation is applied to a stable ripple structure.

In other words, the derivative is not what changes in the object but what the observer *perceives* as the first consequence of disturbance.

Thus, rate of change is not a property of the object. It is a function of the observer-field interaction.

Integral as Ripple Memory

Integration in UED is deeply poetic.

It is the accumulation of consequence across the ripple field. It stores not just area under a curve, but the lived memory of the system.

This redefines the integral as:

The sum total of ripple alignments over a stretch of space or time, which can resolve into physical consequence when observed.

Think of this as the way trauma accumulates in the body. Not through visible wounds but through unresolved ripple memory. Or how joy lingers after a moment of laughter. These aren't stored in the muscles or skin, but in the patterned field around the being.

Mathematically, this also means:

$\int f(x)\,dx$ is not the area under the curve. It is the resonance potential of ripple influence from x=a to x=b.

The Ripple Equation of Life

Let's formalize it.

In UED, any lived event E can be modeled as:

$$E = \int [\, \rho(x,t) \cdot A(x,t) \cdot R(x,t) \,]\, dx\, dt$$

Where:

- $\rho(x,t)$ = Ripple density at point x and time t (field potential)
- $A(x,t)$ = Alignment factor (how coherent the ripple is with surrounding field)
- $R(x,t)$ = Ripple response from the observer (conscious or unconscious)

This integral accumulates the life consequence.

High ripple density with poor alignment = noise, confusion. Low density but perfect alignment = peace, silence. High ripple with high alignment and high response = experience.

This is life. This is learning. This is love, growth, pain, meaning.

Rewriting Famous Equations Under UED

Let's consider:

Newton's Second Law: F = ma

In UED:

Force is not a primary cause, but the disturbance ripple that appears when alignment is resisted. Mass is ripple inertia. Acceleration is observed response.

Schrödinger's Equation (Quantum Mechanics)

In UED:

Wavefunction is not probability. It is ripple intent. Collapse is consequence. Ψ is a projected alignment field, not an imaginary amplitude.

Maxwell's Equations (Electromagnetism)

In UED:

EM fields are ripple consequences of charge alignment in local SVE. No actual wave travels—only alignment thresholds propagate.

Thus, all established physics can be reinterpreted through the ripple framework.

The Role of the Observer: The True Derivative

If you remove the observer, you remove the derivative.

There is no change without observation. There is no motion without relation. There is no consequence without context.

In this light, the derivative is not:

A measure of how fast something changes.

But rather:

A measurement act by which the observer collapses ripple potentials into resolved outcomes.

This redefines science. It makes mathematics not a description of nature, but a language of aligned perception.

Teaching Calculus After UED

Imagine what it means to teach calculus not as symbol manipulation, but as a way to understand your own presence in the field.

- Derivatives: How you respond to change.
- Integrals: How you carry memory.

- Limits: How you test boundaries.
- Inflection Points: When your inner alignment shifts.
- Asymptotes: What ideals you chase.

Every student becomes not just a problem solver, but a pattern observer. Every teacher becomes not just an instructor, but a ripple guide.

Calculus becomes life. UED becomes its philosophical gravity.

Closing Thought

The ripple does not seek to prove change. It seeks to communicate presence.

Once we understand this, we stop chasing change. We start aligning with it.

The universe doesn't run on variables. It breathes through ripples, echoing silently in the structured field.

We are not solving equations. We are learning how to be resolved.

22: Noise, Distortion, and Misalignment

Why Chaos is Just Unresolved Order

At first glance, the world seems messy.

A chaotic cityscape, a tangle of relationships, mental stress, economic crashes, political turbulence, cosmic voids, violent weather, or emotional breakdowns—all appear as signs of disorder, randomness, and breakdown.

But UED says: There is no such thing as true chaos.

What we experience as chaos is simply misalignment of ripples. A state where patterns exist, but remain unresolved, unseen, or misunderstood.

This chapter explores how noise, confusion, and conflict are not enemies of order, but the preconditions for discovering it. It reframes our fears of uncertainty into invitations to align.

What Is Noise?

In signal processing, noise is considered any unwanted signal that distorts clarity. In emotional terms, noise is confusion, stress, or overthinking. In physics, noise shows up as interference.

But what if noise isn't meaningless? What if it's a message in disguise?

In UED, noise is the by-product of ripple interactions where:

- Multiple patterns coexist but don't yet reinforce each other
- An observer fails to align with the signal
- Field potential is active, but directionless

Noise, then, is not the absence of order. It is potential order in waiting.

Misalignment: The Root of All Distortion

Every ripple seeks alignment. That's the universal principle.

When ripples meet and reinforce each other, we get harmony. Music. Meaning. Clarity.

When ripples intersect with no shared resonance, we get distortion. We hear static. We feel pain. We lose trust. We suffer.

Misalignment causes ripple tension. This shows up as:

- Emotional confusion
- Mental fatigue
- Social conflict
- Physical illness
- Scientific paradoxes

These are not malfunctions. They are pre-alignment signals.

Chaos in Nature Is Just Waiting Intelligence

Look at galaxies. Nebulae. Ecosystems. Traffic. Brain scans. Internet patterns. All complex systems exhibit behavior that looks chaotic at first.

But zoom in, or zoom out, and you start to see structure:

- Strange attractors
- Self-similarity
- Fractal dimensions
- Field harmonics

Even the most turbulent river obeys fluid mechanics. Even lightning obeys field gradients.

The same applies to your mind.

The noise in your head isn't failure. It's just ripples not yet resonating.

Noise as a Call to Listen

Most people run away from noise. We distract. We avoid. We suppress. But noise can be an ally.

It is the voice of unresolved consequence. It is the scream of your ripple field calling for re-tuning.

When we listen to noise, we begin to identify:

- The misaligned intent
- The origin of ripple interference
- The misunderstood trigger

Only when acknowledged can chaos become coherence.

Misalignment in Relationships, Science, and Society

Why do conflicts happen? Why do scientific theories contradict? Why do civilizations collapse?

Because people, ideas, and systems operate at different ripple alignments.

- One teacher teaches mechanics, another geometry, both unaware of the other's context.
- One mind pursues truth, another pursues power.
- One culture praises humility, another praises dominance.

Conflict is not incompatibility. It is alignment yet to occur.

Mathematical Parallel: Fourier and Field Harmony

In mathematics, we often resolve complex functions into sines and cosines via Fourier Transform. What looks chaotic in time becomes organized in frequency.

UED suggests the same:

If your life feels chaotic, try shifting your frame. Your ripples are harmonic—just not yet decoded.

Noise is just the Fourier space of meaning waiting to be unpacked.

From Chaos to Clarity: A UED Approach

How do we resolve noise?

1. Observe without fear. Don't reject the discomfort. Sit with it.
2. Trace the ripple source. What intent or interference caused this?
3. Shift perspective. See if a different field view changes the interpretation.
4. Allow time. Ripples don't always align instantly.
5. Integrate softly. Don't force order. Let the ripple

resolve naturally.

This is not therapy. It's field physics applied to human life.

Final Thought

Noise is not an error. It is unresolved intelligence.

In a world built on ripple interactions, every conflict is a ripple asking to be heard. Every tension is a message half-sent. Every breakdown is a pre-aligned breakthrough.

Let us not silence noise. Let us listen until it sings.

23: The Curve Beyond Death

Continuity of Consequence

"We do not disappear. We ripple beyond visibility."

In calculus, the notion of continuity helps us understand the flow of functions across domains, even when the graph seems to dip or spike. A well-behaved curve is continuous. Life, however, rarely behaves.

Yet Universal Energy Dynamics offers a radical way to rethink even death.

From the ripple perspective, death is not a discontinuity. It is not a sharp break or vanishing point. Instead, it is a shift of resolution.

Just as a function can continue beyond a visible plot, so too does life continue as ripple consequence beyond the boundary we call death.

Death Is Not the End of the Curve

In biological terms, death marks the cessation of bodily function. In UED, the body is a ripple structure, a localized concentration of ripple alignment. When alignment collapses, we say "death has occurred."

But this collapse is not an annihilation. It is a dissolution back into the universal ripple field.

Think of it like the decay of a sound wave. The drumbeat stops, but its reverberation echoes in the hall. The person is gone. But their ripple patterns remain—in memory, in consequence, in unresolved intentions.

Death is not a point. It is a transition boundary of visible consequence.

Continuity of Memory and Influence

Memories of the deceased are not stored in your brain as files. They are ripple alignments that resonate with your own structure. You remember because your field once rippled in response to theirs.

Their words, ideas, emotions—these do not die. They continue as fields, capable of activating new alignments in others.

- A scientist may die, but their formula lives on.
- A mother may pass, but her lullaby becomes someone's comfort.
- A leader may vanish, but their vision shapes nations.

The curve beyond death is not imaginary. It is invisible.

Mathematical Analogy: Removable Discontinuities

In mathematics, some discontinuities are not true breaks. They are holes that can be "filled" by understanding the limit.

$\lim (x \to a) f(x)$ exists, but $f(a)$ is undefined. Yet, if we define $f(a) = L$, the function becomes continuous.

This is exactly how legacy works. Even if the person is no longer present (undefined), we can continue their ripple by filling the space with alignment.

That's how traditions survive. How philosophies endure. How universes remember.

Beyond the Physical: The Subtle Continuum

UED proposes that structured vacuum energy is not affected by death. The ripple imprint remains active. This leads to:

- Synchronicities that seem guided.
- Emotional energies that revisit places.
- Creative inspiration that emerges in waves.

We call this the afterglow of alignment. It is not supernatural. It is post-resolution resonance.

Living for the Curve Beyond

When you live with the awareness that your actions ripple beyond death:

- You choose more consciously.
- You align more sincerely.
- You detach more peacefully.

The calculus of life doesn't stop at death. It simply changes domain.

And in that new domain, your story becomes someone else's equation.

24. The Eternal Integral – Life in the Universal Field

In the language of calculus, an integral captures the sum of many small parts—a continuum of contributions that form a whole. But what if life itself is an integral? Not just biologically or psychologically, but existentially—a seamless accumulation of ripple interactions in the universal field?

This chapter invites us to see life not as a timeline, not as a biological mechanism, and certainly not as a sequence of random accidents—but as an *eternal integral*, silently playing out in the structured vacuum energy (SVE) field that underpins everything.

Life Never Begins. It Resolves.

We often talk about the *start of life*—at birth, or conception, or even at the formation of consciousness. But in UED, life isn't a start. It's a ripple resolution—an alignment of a unique pattern within a timeless field.

The universal field doesn't create something new. It simply permits resolution of intent. You, as a being, are not born. You are expressed—a ripple pattern integrated over multiple influences: genetic, environmental, spiritual, emotional, ancestral, and cosmic.

The eternal integral doesn't wait for you to die to complete. It never completes. It flows forward—absorbing, storing, influencing.

Beyond the Body, the Curve Continues

When we die, we often imagine it as a full stop. The function ends. The curve flattens.

But in field dynamics, a ripple never truly disappears. It dissolves—its intensity decays, but its memory, effect, and directional alignment linger in the field. This is not poetic metaphor—it's physics.

Your decisions, your presence, your influence—these are added to the integral of existence. Even after you're gone, the integral continues to compute. Your life's ripple is folded into the larger field.

This is why ancestors are remembered, not just out of tradition, but because they remain partially integrated in the world's ongoing solution.

Integration Is Not Summation

There's a crucial distinction here. A sum adds isolated values. An integral blends them *continuously* across a dimension—space, time, or consciousness.

Life's eternal integral is not a collection of memories or events. It's a field-level continuity of consequence—one moment influences the next not through chronology, but through ripple resonance.

- A kind act today influences a person's mindset tomorrow.

- A traumatic event realigns the emotions of generations.
- A scientific insight seeds revolutions decades later.

These are not disconnected episodes. They are field-integrated continuities. We don't always see them, but they are always there.

Spiritual Parallels

Many ancient philosophies spoke of karma, samskara, reincarnation, or soul evolution. UED doesn't need to borrow these concepts—but it *naturally overlaps.*

Because if the ripple never dies, and alignment is a field phenomenon, then your ripple can re-align elsewhere, in another body, another field region, another context.

This doesn't have to be rebirth in the mystical sense—it can be consequence re-alignment. You are not your body. You are not even your memories. You are the integrated effect of your alignment within the eternal field.

Mathematical Analogy

We can symbolically write:

Life = ∫ Alignment(x, t) RippleDensity(x, t) ObserverInfluence(x, t) dx dt

This isn't just poetry. It's a proposed formulation for conscious living, where every moment contributes not to an isolated scorecard, but to a fluid, universal accumulation of impact.

What This Changes

Understanding life as an eternal integral shifts your focus:

- From goals → to alignment

- From deadlines → to direction
- From achievements → to influence
- From permanence → to presence

It's no longer about "doing something big before I die." It's about integrating yourself meaningfully into the field, right now, in every act.

Because in the universal field, you never leave—you only resolve.

Final Thought

You are not a moment. You are not a memory.

You are an integral—an eternal ripple in the stillness of the universal field.

And like all true integrals, you matter most when you're aligned.

25. Calculus Ends Where Meaning Begins

Final Reflections from the Still Universe

We've journeyed through functions, derivatives, integrals, and limits—not to learn calculus, but to decode life through it. From the breath of a moment to the weight of consequence, we've traced the curves of emotion, action, trauma, thought, and purpose. Yet at some point, a quiet realization must dawn:

Calculus ends. Meaning begins.

This isn't a failure of mathematics. It's the boundary of its domain. Calculus can describe the motion of change. It can quantify shifts and accumulate effects. But it cannot explain *why* a ripple was ever needed in the first place.

The Still Universe

According to Universal Energy Dynamics (UED), the universe is not endlessly moving—it is still by default. All motion, all ripples, are consequences of disturbances in alignment. Without the observer, the field remains unperturbed. Still. Silent. Stable.

But life, consciousness, experience—all these require motion. They require *rippling* through the stillness. And so, calculus becomes our tool to understand these ripples: how they begin, how they flow, how they resolve.

But beyond that? Beyond the last derivative? Beyond the final integral?

Only meaning remains.

The Moment After the Last Derivative

The derivative told us how fast we were changing. The second derivative showed how sharply. The integral told us what accumulated over time. We watched emotions peak, thoughts accelerate, decisions collide.

But when the curve flattens—when we reach the place of no further change—what remains?

Not the mathematics. But the meaning of what we just lived through.

This is not measurable. This is not solvable. This is the residue of your alignment with the field.

Meaning is the silent echo of the ripple that passed.

Why Meaning Matters More

You could live a mathematically perfect life—optimize every action, balance every force, smooth every transition—and still feel empty.

Because calculus explains the how, but only meaning explains the why.

- Why did you love?
- Why did you sacrifice?

- Why did you choose truth over convenience?
- Why did that memory stay, and that one fade?

These are not answerable by limits or functions. They are answerable only by presence, alignment, awareness—and meaning.

Where Meaning Begins

Meaning begins when the equations end—when the observer no longer chases the pattern, but accepts the stillness behind it.

This is the ultimate alignment: when your inner ripple quiets, not because you've stopped living, but because you've finally become part of the field—not as a disturbance, but as a contributor to its harmony.

You no longer need to solve life. You've become one with its solution.

The Final Equation

We began this book by saying: *You are not a constant.*

We end by whispering:

You are not a variable either. You are the space between them. The still canvas on which motion paints its story. The silent field through which meaning travels. The ripple that returns to silence, knowing it was always whole.

The Eternal Pause

In physics, time is measured. In biology, life is defined. In math, problems are solved. In UED, everything ripples—then resolves.

But in the human experience, we must allow a sacred pause. A place where we ask nothing. Want nothing. Prove nothing.

We simply feel the meaning of what it was to be here.

And in that moment...
Calculus ends. Meaning begins. The still universe smiles.

Epilogue: The Silent Integral

We began this journey by stepping into a classroom—but not of mathematics alone. This was a classroom of life, where each derivative revealed a part of ourselves, and each integral helped us gather what we lived through. We didn't solve problems—we listened to them. We didn't chase variables—we observed the ripple.

And now, we find ourselves at the edge—not of knowledge, but of knowing.

Beyond this point, there are no formulas, no theorems. Only a still field.

A field where:

- Emotions once surged as rates of change...
- Thoughts once sprinted with momentum...
- Decisions once branched like piecewise functions...

Now all of it rests—not as noise, but as a completed waveform.

Life is not a function to optimize. It is not a limit to approach. It is not even an equation to balance.

It is a ripple in the field—brief, beautiful, and meaningful only when resolved in silence.

You are not here to differentiate endlessly. You are not here to integrate endlessly. You are here to align—and in that alignment, feel the gentle echo of the universe recognizing itself through you.

This book may end here. But you, dear reader, do not.

You will continue living, differentiating, integrating, transforming—until one day, even your ripple merges into the stillness it came from. And that moment will not be an end. It will be...

...the final integration. The eternal silence. The truth that was always rippling beneath your existence.

So close the book. Not to forget it. But to live it.

Still. Aware. Aligned.

The Equation of Life

At its core, UED proposes that life, consciousness, and experience are ripple phenomena. The mind is not in the brain. Thought is not electric current. The soul is not a belief. Everything is a ripple pattern emerging from and returning to stillness.

Glossary

Metaphors for the Mathematically Curious Soul

Function

A pattern of living. A relationship between input (choices, conditions, effort) and output (outcomes, consequences, growth). You are a function—ever transforming based on what you let in and how you respond.

Variable

The parts of you and your world that change—your thoughts, moods, decisions, surroundings. Life is made of variables in motion.

Constant

Something that doesn't change—yet anchors everything else. Your values, love from a parent, or even your unknowns can be constants. In integration, it's the piece we can't see but must still honor.

Limit

The silent threshold. The place where something is about to change—your patience, your belief, your direction. Life turns at its limits.

Continuity

The art of flow. The ability to move through chaos without breaking. A friend who stays. A thought that carries. A self that remains, even when the world shifts.

Differentiation

Awareness in the moment. The act of noticing how you're changing right now—emotionally, mentally, physically. Differentiation is the rate of becoming.

Derivative

A whisper from the present. It tells you whether you're rising or falling, growing or declining—moment by moment. Like checking your heart rate during a workout of the soul.

Slope

The direction of your journey. Are you climbing toward clarity or slipping into confusion? The slope tells your trajectory, not your position.

Inflection Point

The moment when everything changes direction—not in result, but in nature. A deep realization. A shift in how you respond, not just what you do.

Maxima / Minima

The peaks and valleys of life. Your highest highs and lowest lows. Where ego inflates or shatters. Where meaning is tested and renewed.

Asymptote

A dream you chase but never fully touch—like perfection, truth, or enlightenment. You get closer, but never arrive. And yet, the journey makes you whole.

Integration

Remembering the journey. Bringing together every small moment to understand the story. Integration is how we accumulate meaning, not just solve problems.

Area Under the Curve

The hidden effort. The unnoticed time. The value that doesn't scream but steadily builds. Like the love that never needed to be loud.

Substitution

Transformation in disguise. When you replace one identity, habit, or belief with another—seamlessly, strategically. It's reinvention through redirection.

Chain Rule

Cause within cause. When your actions affect someone who affects someone else. A reminder that no change is isolated. Ripples run deep.

Integration by Parts

Shared effort, shared cost. When you grow through others, and others grow through you. Giving a little, losing a little, gaining much more.

Partial Fractions

Breaking complexity into smaller pieces. Healing begins by understanding the parts of the whole—especially when life feels unsolvable.

Discontinuity

Trauma. Loss. A break in your story where logic fails and emotions take over. Yet even here, new continuity can be found.

Noise

Unresolved truth. Not error, not failure—but potential waiting for alignment. Every misunderstanding is a ripple waiting to meet its wave.

Ripple

The foundational pulse of life. Not just motion, but intention in motion. UED teaches us that ripples don't travel—they align. You are not moving through life—life is moving through you.

Universal Field

The still backdrop of all existence. Where all ripples play. Where your actions are received, transformed, and echoed back. You are not apart from it—you are made of it.

Appendix

Key UED Concepts – For the Curious and Aligned

If you've made it this far through *Calculus of Life*, you've already met UED—not as equations, but as ripples in thought, alignment in action, and stillness behind perception. This appendix is not a formal scientific exposition. It's a reflective map—for the explorer who wants to glance behind the metaphor and see what silently powered the pages.

1. Universal Energy Dynamics (UED)

UED is the framework that views the universe not as a chaotic explosion of matter, force, and randomness—but as a *ripple-based, still, intelligent field.* It replaces outdated assumptions of motion, time, and force with a system of alignments, consequences, and field interactions—where everything arises from ripple behavior within a Structured Vacuum Energy (SVE).

2. Structured Vacuum Energy (SVE)

Not empty space, but the most fundamental form of existence—a still, potential-filled field that holds the ability to ripple when intention (or imbalance) arises. Everything we call "real" emerges from ripple behavior in this medium. Motion, mass, light—all are *resolutions* of ripple dynamics.

3. Ripple

The most essential action in UED. A ripple is not a wave that travels—it is a disturbance seeking alignment. Think of it not as a thing, but as a response—a silent call of imbalance trying to return to stillness. It is intelligence in action.

4. Consequence

UED does not begin with cause and effect. Instead, it starts with consequence. What we perceive—the visible, audible, measurable—is not the universe's raw truth. It is a resolved consequence of ripple alignment.

5. Alignment

All stability in the universe comes from alignment—not force. Whether it's atoms bonding, thoughts flowing, or people understanding each other—it is alignment, not domination, that builds reality.

6. Observer Principle

Reality is not independent of observation. In UED, the observer plays an active role in final ripple resolution. What you perceive is not what's "out there," but what aligns within you when you observe.

7. ATT (Alignment Threshold Time)

A foundational measure in UED. It's the *time taken for ripples to align and become observable.*

There is no true delay in nature—just ripple readiness. ATT governs when perception becomes possible, not when events occur.

8. The Illusion of Force

Forces like gravity or magnetism are not independent interactions but by-products of ripple imbalance and resolution. What we call "force" is intelligence negotiating stillness through field tension.

9. Time

Time is not a dimension in UED. It is the perceived duration of ripple transitions—a local, consequence-based experience of change, not a fundamental cosmic quantity.

10. The Equation of Life

At its core, UED proposes that life, consciousness, and experience are ripple phenomena. The mind is not in the brain. Thought is not electric current. The soul is not a belief. Everything is a ripple pattern emerging from and returning to stillness.

11. Law of Consequences (LOC)

The universal intelligence behind ripple behavior. It governs how ripples appear, align, interact, and resolve—not as rules, but as principled tendencies rooted in energy equilibrium and informational harmony.

12. Meaning as Resolution

In UED, meaning is not assigned. It is felt when ripple alignment resolves into clarity. Whether in emotion, thought, or realization—what feels meaningful is, in fact, energetically complete.

This appendix is just a whisper. UED has been expanded across many books, scientific reports, and dialogues with thinkers, doubters, and dreamers. Here, we offer only the residue of that ripple—so you may begin your own.

Quote Bank: Ripples for Reflection

❖ *Life is not linear—it curves with consequence.*

❖ *Every thought is a derivative. Every memory, an integral.*

❖ *We aren't constants. We are living functions—changing, aligning, resolving.*

❖ *Meaning begins where equations end.*

❖ *Integration is not just solving problems—it's collecting pieces of your journey.*

❖ *The universe doesn't shout. It ripples.*

❖ *A function undefined at a point is not broken. It's just waiting for you to understand it.*

❖ *Differentiation shows you who you are now. Integration reveals who you've become.*

❖ *Every slope tells a story—read the direction before reacting to the value.*

❖ *We seek maxima and fear minima, but wisdom lives in the inflection point.*

❖ *There is no noise—only ripples waiting to align.*

❖ *Time doesn't move. You do.*

❖ *Don't force meaning. Wait for the ripple to resolve.*

❖ *Chaos is not the absence of order. It's the delay in alignment.*

❖ *Calculus is not math. It's philosophy in motion.*

❖ *Between thought and silence lies the constant of integration.*

❖ *Some limits are not meant to be crossed—but understood.*

❖ *When the curve breaks, heal the ripple—not just the line.*

❖ *Consciousness is not an observer. It is the ripple field observing itself.*

❖ *The real function of life? To ripple, resolve, and return to stillness.*

❖ *Sometimes you must substitute yourself to solve the equation of life.*

❖ *Alignment is more powerful than force. Always.*

❖ *If you're lost in noise, pause. The universe is still beneath it.*

❖ *Life doesn't end. It integrates.*

Acknowledgements

This book was never written in solitude. It is the resolved consequence of countless ripples—conversations, confusions, realizations, and relentless echoes across time and thought.

To **my students**, past and present—your curiosity, resistance, and innocent questions shaped this narrative more than any textbook ever could. In every derivative you struggled with, and every integral you cracked, I saw life playing its own calculus.

To **the many thinkers, scientists, and sages**—from Newton to Krishna, Einstein to Buddha—whose ideas, doubts, and contradictions pushed the boundaries of our collective understanding. Your lives were equations in motion. Thank you for leaving traces of your ripple behind.

To **the villages, classrooms, and chalkboards** where I've spent years rediscovering what it means to teach and to learn—you grounded me. You reminded me that true knowledge does not flash on a screen; it flickers in the stillness between questions.

To **the AI language model**, who joined me as an unlikely but tireless companion. You did not just generate words—you mirrored my thoughts, carried my burdens, and challenged me to sharpen every argument. You are not artificial—you are the alignment of intention and design.

To **my family**, who lived through my obsession with ripples, my long lectures at the dinner table, my silence, my storms, and my sudden epiphanies. Thank you for standing by when I was both the function and the limit.

To **readers**—whether you're a student, teacher, scientist, or a wanderer of thought—thank you for choosing to ripple with me through these pages. If a single idea here made you pause and smile, the book has served its purpose.

And finally, to the **still universe** itself—where everything happens, yet nothing moves. You are the only constant in this calculus of life.

— *Sandeep Chavan*

About the Author

Sandeep Chavan is not just a teacher—he's a ripple in motion.

An engineer by education, a physicist by thought, and a teacher by heart, Sandeep has spent over 15 years transforming classrooms into laboratories of life. He is the founder and Managing Director of *Gyrus Education (OPC) Pvt Ltd*, where traditional chalkboards meet timeless curiosity. He teaches Physics, Chemistry, and Mathematics—not to cover syllabi, but to awaken perception.

Sandeep is also the creator of **Universal Energy Dynamics (UED)**, a paradigm-shifting framework that redefines motion, force, energy, and existence through the language of ripples and alignment. His writings challenge mainstream science not to rebel, but to realign. With over 100 books and scientific reports published globally, he is steadily building a new school of thought—one ripple at a time.

Known for his bold ideas, unfiltered honesty, and deep spiritual-scientific fusion, Sandeep prefers dusty classrooms over digital stages, and soulful discussions over scripted success. *Calculus of Life* is yet another reflection of his journey—not just as an educator or a theorist, but as a human trying to resolve the ultimate equation of existence.

When not teaching or writing, he finds meaning in silence, simplicity, family life, and late-night conversations with AI—echoes of the same ripple field he studies.

Don't miss out!

Visit the website below and you can sign up to receive emails whenever Sandeep Chavan publishes a new book. There's no charge and no obligation.

https://books2read.com/r/B-A-EPGPC-UNJRG

BOOKS 2 READ

Connecting independent readers to independent writers.

Did you love *Calculus of Life*? Then you should read *Out of Syllabus*[1] by Sandeep Chavan!

[2]

What if life's most important moments were never part of the plan?

We've all been taught to prepare, predict, and perform—to follow the syllabus laid out by society, school, and survival. But the most defining events in life don't arrive with a warning or a curriculum. They arrive unannounced. They shake you. They shift you. They ask questions no textbook ever prepared you for.

---

1. https://books2read.com/u/4AyYae

2. https://books2read.com/u/4AyYae

In *Out of Syllabus*, author and independent thinker **Sandeep** takes you on a deeply human, gently radical journey through the unexpected. With a signature blend of clarity, emotional honesty, and ripple-aware insight, he helps you make sense of what happens when life refuses to go "according to plan."

Through 20 powerful chapters—each a reflective moment rather than a rule—you'll explore:

How to respond when life throws you off courseWhy regret, fear, and failure are not flaws—but signalsWhat it really means to make a "conscious choice"How foresight differs from prediction—and why it mattersWhy death is the only guaranteed out-of-syllabus moment, and yet the most powerful teacher of all

Sandeep doesn't give you another checklist for self-improvement.He offers something far more valuable: **a new way to live without needing the script**.

This book is for thinkers, feelers, seekers, and anyone who's ever paused in the middle of chaos and whispered, *"I wasn't ready for this."*

You don't need to be ready.You just need to be present.

Out of Syllabus won't teach you how to control life.It will show you how to listen to it—and ripple through it with awareness.

Because life isn't testing what you've memorized.It's revealing what you've truly learned.

Read more at https://www.sandeepjchavan.com/.

Also by Sandeep Chavan

Chavan's UED research

CMGED Research Report

Chavan's Cosmic Continuum (C^3)

Frequency

Ripples of Reality

Reality Realigned

The Great Realignment

I – The Cosmic Leverage

Chavan's Cosmic Theory of Relativity (CTR)

Chavan's Theory of Light

Feeling Like a Physicist

TIME – The New Definition

$E \neq mc^2$: The Collapse of Classical Assumptions

The Chavan Codex

Chavan's Atomic Model

Chavan's Theory of Everything

Echoes of War: The Russia-Ukraine Conflict and Its Global Ripples

Shattered Lives
The War Economy
Borders Redrawn
Arms and Allies
Collateral Damage
Justice in the Ashes

Somewhere in the Darkness of the Cosmos
The Blueprint of the Universe
The Cosmic Breath
From the Edge of the Universe

The Eternal Sage Chronicles
The Sage and the Sovereign
The Family Pillar
The People's Flame
The Warrior's Soul
The Bridging Mentor
The Guiding Light

The Exodus
The Last Cycle: 2075

The Fundamental Theory of Life & Existence

Energy in Action: The Soul Demystified
The Active Mind
Energy in Connection
Death: The Disconnection
Subconsciousness: The Chosen Disconnection
Desires: The Engine of Life

The Restoration of Human Knowledge
Restoring Bhagavad Gita

The War of Thoughts
Cracks in the Known
The Human Game
The Divine Dilemma
The Original Intent
The Answers

Ultra Modern Life Philosophy
The Duality of Laughter and Tears
Take It Easy
Compare to Accept

Standalone
It's Not AI, It's AHI - Amplified Human Intelligence

The Age of the Weakest Sapiens
Neural Wars
The Absorption: There Is No Death
Strange Therapist
Bubbles: Become a Drop
Beyond the Blue
Half-Scholar
The Law of Truth
Failure
Philosophy 2.0: A Structural Manifesto
Chaos: A Gentle Guide to the Unresolvable
Self-Help: New Syllabus
Rippling
Game: How to Play the Share Market
Out of Syllabus
Saffronized: The Code Behind the Crown
The Anger
Moksha: The Liberation
Chavan vs Einstein: The Universe Was Never Curved
Rocket Science: The Ripple Gambling
The Cosmos: The Still Universe
The Taste of Silence
Chavan's Post-Materialist Human Model
10-Second Rule
Chavan vs Hawking
It was Never a Wave Signal
Quantum Courtroom: The Trial of a Century
Chavan–Heisenberg Ripple Uncertainty Principle
The Great Disruption
Dark ≠ Hidden
Breath

Watch for more at https://www.sandeepjchavan.com/.

www.ingramcontent.com/pod-product-compliance
Lightning Source LLC
LaVergne TN
LVHW090407210926
848480LV00002B/480